AF576020

Solid Fuels Technology and Applications

Solid Fuels Technology and Applications

Editors

Nikolaos Koukouzas
Pavlos Tyrologou
Petros Koutsovitis

MDPI • Basel • Beijing • Wuhan • Barcelona • Belgrade • Manchester • Tokyo • Cluj • Tianjin

Editors
Nikolaos Koukouzas
Centre for Research & Technology Hellas/Chemical Process and Energy Resources Institute (CERTH/CPERI)
Greece

Pavlos Tyrologou
Center for Research & Technology Hellas/Chemical Process and Energy Resources Institute (CERTH/CPERI)
Greece

Petros Koutsovitis
Department of Geology, University of Patras, Patras
Greece

Editorial Office
MDPI
St. Alban-Anlage 66
4052 Basel, Switzerland

This is a reprint of articles from the Special Issue published online in the open access journal *Energies* (ISSN 1996-1073) (available at: https://www.mdpi.com/journal/energies/special_issues/SFTA).

For citation purposes, cite each article independently as indicated on the article page online and as indicated below:

LastName, A.A.; LastName, B.B.; LastName, C.C. Article Title. *Journal Name* **Year**, *Volume Number*, Page Range.

ISBN 978-3-0365-0322-6 (Hbk)
ISBN 978-3-0365-0323-3 (PDF)

Contents

About the Editors

Nikolaos Koukouzas holds a BSc degree in Geology with MSc and PhD degrees in Industrial Mineralogy from the University of Leicester, UK. He received his bachelor's degree in Geology from the National University of Athens in 1987. He continued his postgraduate studies in the UK, receiving a 4-year grant from the State Scholar-ship Foundation of Greece, and completed his PhD in 1995. Dr. Koukouzas is Director of Research at the Centre for Research and Technology Hellas (CERTH) and Director of the Laboratory of the Solid Fuels at CERTH, supervising a team of scientists with offices in Athens, Thessaloniki, and Ptolemais, Greece. He is Member of the Executive Committee of the Monitoring and Assessment Exercise of the Research Fund for Coal and Steel, National Delegate in the Government Group of Zero Emission Power Plant Technology Platform (ETP-ZEP), and National Delegate in the Policy and Technical Group of the Carbon Sequestration Leadership Forum (CSL) coordinated by the US DoE. His research expertise is focused on coal geology, industrial mineralogy, coal geochemistry, fossil fuel characterization, coal combustion by-products, coal slope stability, waste and mine water treatment, and CO_2 storage and monitoring. Dr. Koukouzas has been Project Manager and had scientific responsibility of several re-search projects with over 25 years' experience on energy and environment issues. He has 250 publica-tions, 1650 citations, and an h-index of 21.

Pavlos Tyrologou is a research fellow at CERTH, holds a BSc (4 years Hons) degree in Geology and Applied (engineering) Geology from Glasgow University, an MSc in Applied Environmental Geology from Cardiff University, and a PhD in Environmental Geotechnical Engineering from Imperial College London, fully supported by a scholarship award from The Institute of Materials, Minerals and Mining, UK. He is professionally accredited in the UK both as a Chartered Geologist and Eurogeologist in the field of Engineering Geolo-gy. Pavlos has 15 years of consultancy experience in various public infrastructure design/works in Greece, United Kingdom, Madagascar, and Sint Maarten. He is an expert witness for both the Athens Public Prosecutor's Office and Athens Court of Justice. He has evaluated more than 65 research and business projects as an independent evaluator for both the European Commission and Greek Govern-ment. Due to his expertise, Pavlos serves as Coordinator in the European Federation of Geologists (EFG) on the panel Natural Hazard and Climate Change and is member of the panels of experts on soil pro-tection. On the same note, he is the Greek representative within the EFG council through the Associa-tion of Greek Geologists. He is a member of the Greek Society for Soil Mechanics and Geotechnical En-gineering, Geological Society of London, International Society for Rock Mechanics, and Member of In-ternational Society for Soil Mechanics and Geotechnical Engineering, at which he serves on the tech-nical committee of the Engineering Practice of Risk Assessment and Management. He also has exten-sive research experience in several European H2020 projects, including proposal writing, inception, and execution (CHPM30, UNEXMIN, SLOPES, KINDRA, INTRAW, COALTECH2051, StrategyCCUS, LEILAC2, RECPP) of various tasks such as project management and research activities. His current re-search interests are focused on climate change and related activities and how these are related to securi-ty issues such as state fragility and vulnerability.

Petros Koutsovitis is Assistant Professor at the University of Patras, Department of Geology. He holds a PhD in the fields of Mineralogy, Petrology and Geochemistry from the University of Athens, Greece. Dr. Koutsovitis has been a Post-doctoral researcher at the University of Vienna (under an FWF-funded project) and was later employed at the Greek Institute of Geology and Mineral Exploration, within the frame of ESTMAP (Horizon 2020) and YPOTHER (NSRF-funded) projects. He continued his research activities at the Centre for Research & Technology (CERTH), participating in EU-funded Horizon2020 and RFCS project's (STRATEGY-CCUS, COAL2GAS & COALBYPRO), as well as at the University of Athens as an Academic Educational Fellow. Dr. Koutsovitis has also participated in the INTRAW Horizon 2020 programme, in cooperation with the European Federation of Geologists (EFG). He has published several papers in peer reviewed Journals and has been awarded by the Academy of Athens for his research.

Preface to "Solid Fuels Technology and Applications"

We are delighted to deliver the "Solid Fuels Technology and Applications" Special Issue of Energies. This Special Issue aims to promote research and technological development on the integration and exploitation of solid fuels and their by-products. Solid fuels are still a key factor in energy production, having been enhanced through recent developments. Under the current trend of climate change and the necessity to mitigate greenhouse gas emissions, high efficiency is more than ever a prerequisite for low financial and environmental cost production.

The seven papers selected in the current Special Issue present novel applications that bring forward the current state-of-the-art on solid fuels utilization. Multi-faceted solutions to a challenging environment are delivered to the readers of this unique collection of articles that combine innovative applications to emerging problems and needs.

In this Special Issue, O'Brien et al. [1] developed a research model that addresses both the sustainability and the profitability of bioethanol production due to synergy with the torrefaction of DDGS and using produced biochar as a marketable fuel. Their results clearly demonstrated the relationship between reduction in environmental footprint (24% reduction in CO_2 emissions) and the introduction of comprehensive on-site valorization of dried distillers' grains with solubles (DDGS). Wu et al. [2] found that the increased OH/H ratio in staged O_2/CO_2 combustion offsets part of their reproducibility, resulting in the final NOx emissions being higher than those in air combustion under similar conditions.

Where are the potential sites for underground energy and CO_2 storage in Greece? Arvanitis et al. [3] provided a geological and petrological research approach that can serve as a basis for future research and deployment of promising areas. The use of mechanically activated serpentinite was considered by Petrounias et al. [4] to remove Cu(II) from industrial wastewaters. It was proposed that the higher mechanical activation of the studied serpentinites (being subjected to a 1500 revolutions LA test) is related to their higher capability of performing Cu removal. Chećko et al. [5] presented an assessment of the resources of methane, considered as the main phase in the prospective areas of the Upper Silesian Coal Basin, Poland. The results of the numerical simulations confirm that the application of multilateral well systems combined with hydraulic fracturing considerably improves the efficiency of CBM extraction from seams characterized by low coal permeability. Nazos et al. [6] dealt with the treatment of barley straw by acid-catalyzed wet torrefaction (ACWT) in a Parr 4553 3.75 L batch reactor (autoclave). The findings indicated that the composition changes of the straw due to ACWT had a significant effect on the HHV of the pretreated material. The petrographic features of sandstones in Central Greece were examined by Petrounias et al. [7] to evaluate their behavior in construction and in energy storage applications. Petrographic methodologies were combined with the quantification of modal composition (GIS proposed method) and 3D depictions of their petrographic features (3D Builder software).

The guest editors are confident that readers will enjoy the articles presented in this beneficiary Special Issue of "Solid Fuels Technology and Applications".

Nikolaos Koukouzas, Pavlos Tyrologou, Petros Koutsovitis

Editors

References

1. O'Brien, S.; Koziel, J.A.; Banik, C.; Białowiec, A. Synergy of Thermochemical Treatment of Dried Distillers Grains with Solubles with Bioethanol Production for Increased Sustainability and Profitability. *Energies* **2020**, *13*, 4528.
2. Wu, S.; Che, D.; Wang, Z.; Su, X. NOx Emissions and Nitrogen Fate at High Temperatures in Staged Combustion. *Energies* **2020**, *13*, 3557.
3. Arvanitis, A.; Koutsovitis, P.; Koukouzas, N.; Tyrologou, P.; Karapanos, D.; Karkalis, C.; Pomonis, P. Potential Sites for Underground Energy and CO_2 Storage in Greece: A Geolog-ical and Petrological Approach. *Energies* **2020**, *13*, 2707.
4. Petrounias, P.; Rogkala, A.; Giannakopoulou, P.P.; Lampropoulou, P.; Koutsovitis, P.; Kou-kouzas, N.; Laskaris, N.; Pomonis, P.; Hatzipanagiotou, K. Removal of Cu (II) from Industrial Wastewater Using Mechanically Activated Serpentinite. *Energies* **2020**, *13*, 2228.
5. Chećko, J.; Urych, T.; Magdziarczyk, M.; Smoliński, A. Resource Assessment and Numeri-cal Modeling of CBM Extraction in the Upper Silesian Coal Basin, Poland. *Energies* **2020**, *13*, 2153.
6. Nazos, A.; Grammelis, P.; Sakellis, E.; Sidiras, D. Acid-Catalyzed Wet Torrefaction for En-hancing the Heating Value of Barley Straw. *Energies* **2020**, *13*, 1693.
7. Petrounias, P.; Giannakopoulou, P.P.; Rogkala, A.; Kalpogiannaki, M.; Koutsovitis, P.; Damoulianou, M.-E.; Koukouzas, N. Petrographic Characteristics of Sandstones as a Basis to Evaluate Their Suitability in Construction and Energy Storage Applications. A Case Study from Klepa Nafpaktias (Central Western Greece). *Energies* **2020**, *13*, 1119.

Nikolaos Koukouzas, Pavlos Tyrologou, Petros Koutsovitis
Editors

Article

Potential Sites for Underground Energy and CO_2 Storage in Greece: A Geological and Petrological Approach

Apostolos Arvanitis [1], Petros Koutsovitis [2,*], Nikolaos Koukouzas [3], Pavlos Tyrologou [3], Dimitris Karapanos [3], Christos Karkalis [3,4] and Panagiotis Pomonis [4]

1 Hellenic Survey of Geology and Mineral Exploration (HSGME), 13677 Attica, Greece; arvanitis@igme.gr
2 Section of Earth Materials, Department of Geology, University of Patras, GR-265 00 Patras, Greece
3 Centre for Research and Technology, Hellas (CERTH), 15125 Marousi, Greece; koukouzas@certh.gr (N.K.); tyrologou@certh.gr (P.T.); karapanos@certh.gr (D.K.); karkalis@certh.gr or chriskark@geol.uoa.gr (C.K.)
4 Department of Mineralogy and Petrology, Faculty of Geology and Geoenvironment, National and Kapodistrian University of Athens, Zografou, P.C. 15784 Athens, Greece; ppomonis@geol.uoa.gr
* Correspondence: pkoutsovitis@upatras.gr; Tel.: +30-26-1099-7598

Received: 9 April 2020; Accepted: 25 May 2020; Published: 28 May 2020

Abstract: Underground geological energy and CO_2 storage contribute to mitigation of anthropogenic greenhouse-gas emissions and climate change effects. The present study aims to present specific underground energy and CO_2 storage sites in Greece. Thermal capacity calculations from twenty-two studied aquifers (4×10^{-4}–25×10^{-3} MJ) indicate that those of Mesohellenic Trough (Northwest Greece), Western Thessaloniki basin and Botsara flysch (Northwestern Greece) exhibit the best performance. Heat capacity was investigated in fourteen aquifers (throughout North and South Greece) and three abandoned mines of Central Greece. Results indicate that aquifers present higher average total heat energy values (up to ~6.05×10^6 $MWh_{(th)}$), whereas abandoned mines present significantly higher average area heat energy contents (up to ~5.44×10^6 $MWh_{(th)}$). Estimations indicate that the Sappes, Serres and Komotini aquifers could cover the space heating energy consumption of East Macedonia-Thrace region. Underground gas storage was investigated in eight aquifers, four gas fields and three evaporite sites. Results indicate that Prinos and South Kavala gas fields (North Greece) could cover the electricity needs of households in East Macedonia and Thrace regions. Hydrogen storage capacity of Corfu and Kefalonia islands is 53,200 $MWh_{(e)}$. These values could cover the electricity needs of 6770 households in the Ionian islands. Petrographical and mineralogical studies of sandstone samples from the Mesohellenic Trough and Volos basalts (Central Greece) indicate that they could serve as potential sites for CO_2 storage.

Keywords: underground; energy storage; natural gas; carbon storage; hydrogen; thermal energy; CO_2

1. Introduction

The use of fossil fuels as energy sources is one of the major contributors of anthropogenic greenhouse gas emissions that include CO_2 [1–3]. To mitigate the effects of global warming the implementation of CO_2 Capture and Storage (CCS) practices is considered as a state-of-the-art technology that aims to reduce emissions of CO_2 into our living atmosphere [4]. To achieve efficient and sustainable energy management it is important to promote practices that aim to reduce the carbon footprint through utilisation of renewable energy resources (wind, biomass, solar and geothermal energy) in Greece with integration of energy storage concepts [5]. This can be achieved with gradual transition to eco-friendly energy systems that promote energy production with the inclusion of Renewable Energy Sources utilisation. This energy transition presents significant and important developments, such as the geothermal energy exploitation in Kenya [6]. However, their global contribution in total renewable energy production is small. This is attributed to the obvious difficulties that are linked with the supply of renewable sources. In the current stage, even more countries tend to adapt and promote the use and research on renewable energy sources [7].

More specifically, energy storage applications as a concept aim to provide technologies that convert energy into storable forms [8]. It also balances energy consumption with production by storing excess energy for long and/or short periods [9]. Despite the cover of energy demands (i.e., seasonal, daily), energy storage provides additional benefits such as: decrease of operational costs, integration of variable energy sources (such as wind, solar, natural gas and geothermal energy) and reduction of environmental effects (low carbon energy supply). Supplementary energy conversion processes are often required, depending on the source type and the implemented storage technology. Energy storage systems can be distinguished into mechanical, chemical, biological, magnetic, thermal and thermochemical types [10]. The choice of the appropriate storage technology depends on the storage purpose, the type of energy source the available storage capacity cost, lifetime and environmental impact [8]. Energy storage systems are distinguished into ground (ground-based ex-situ) and underground (geological in-situ) types. Underground energy storage requires suitable geological reservoirs such as: depleted hydrocarbon reservoirs, rock bodies with appropriate lithotypes (ultramafic rocks, basalts and sandstones), deep saline aquifers, salt caverns and abandoned mines.

Underground Thermal energy storage systems (UTES) provide the opportunity to store thermal energy, by utilising the heat capacity of underground soil and/or rock volumes [11]. To date, the best-established types of seasonal UTES systems [12,13] are aquifer storage (ATES), borehole storage (BTES), cavern storage (CTES), pit storage (PTES) and seasonal tank storage (TTES). There are many specific localities with potential for underground energy storage in Greece [14]. Aquifers for seasonal underground heat or natural gas storage are in sedimentary basins of the Greek mainland (such as Thessaloniki basin; North Greece), as well as in the Aegean islands (such as Evia, Lesvos, Chios and Rhodes). Natural gas can be sufficiently stored in depleted natural gas reservoirs that exist in Greece (Figure 1), which are selected on the basis of a sufficiently large volume of pore space, preferably high permeability of reservoir rocks and the absence of gas admixtures such as hydrogen sulphides. Hydrocarbon reservoirs that could serve as potential sites for underground gas storage include those of Epanomi (Thessaloniki, North Greece), Katakolo (South Greece) and Prinos (North Greece) [14].

Apart from the aforementioned, hydrogen is considered nowadays as an important key source for clean, secure and affordable energy, since it can improve efficiency of power systems and reduce environmental impact of power production [15]. Hydrogen can be stored into porous reservoir rocks such as depleted natural gas or petroleum deposits [16] and salt caverns [16,17]. In Greece, the most suitable formations for considering hydrogen storage (Figure 1) are the caverns in salt/evaporite formations due to the fact that they allow higher injection and withdrawal rates compared to other means of storage [18] thus resulting in high energy deliverability to the energy grid. Evaporite formations of Heraklion (Crete Island; South Aegean) are suitable for underground gas storage, whereas Corfu and Kefallonia Islands (Ionian Sea) could serve as sites for underground hydrogen storage. The abandoned mines of Mandra and Chaidari regions (Attica; Central Greece), as well as the Aliveri mine (Evia Island; Central Greece) could be exploited for underground heat storage purposes.

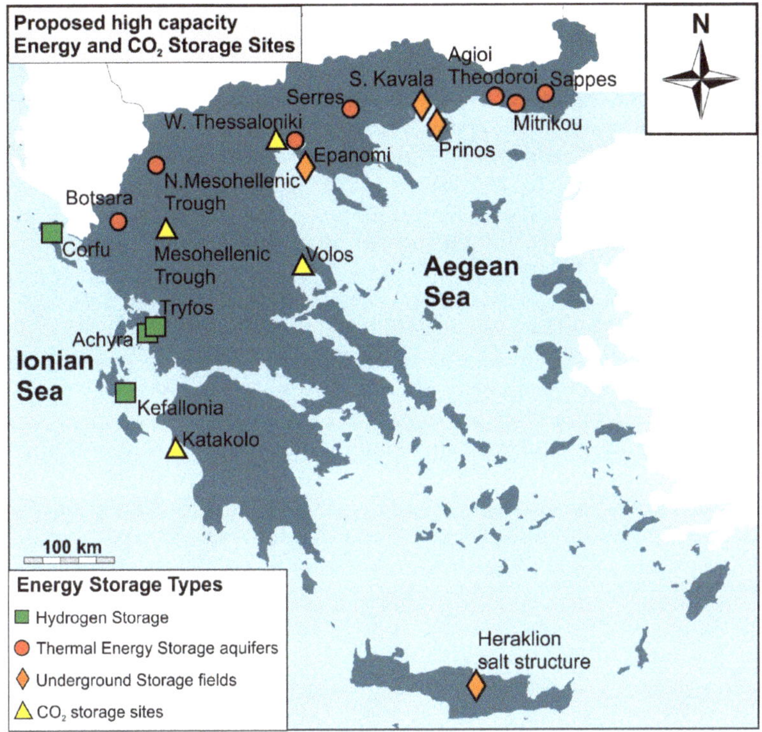

Figure 1. Main proposed high capacity energy and CO_2 storage sites (hydrogen storage: 26,600 $MWh_{[e]}$; Thermal Capacity: 2930–4651 MJ; Gas Storage Capacity: 928,097–4,826,105 $MWh_{[e]}$; CO_2 storage: 27,600 t [3] to 1350 Mt [19].

Several research studies have developed effective methods for carbon capture, by utilising technologies such as membranes, adsorption-based separation of CO_2 [20]. There are many options for Geological CO_2 storage such as deep saline formations [21], abandoned coal mines [22], salt caverns [23], coal seams [24] and depleted hydrocarbon fields [25]. In many cases, CO_2 storage can be combined with extraction of crude oil (CO_2-Enhacned Oil Recovery [26]) or natural gas from hydrocarbon reservoirs (CO_2-Enhanced Gas Recovery). Mineralisation is an alternative option of CO_2 sequestration. Several rock types have been considered as potential underground reservoirs for mineral storage of carbon, or to withhold for a relatively short period of time specific amounts of CO_2. These include: (a) basalts with high porosity, silica undersaturated nature, abundance of plagioclase and feldspars and low alteration grade (recent age) [3,27,28], (b) sandstones [29,30] and (c) serpentinite bodies [31,32].

The most common parameters that must be considered prior to the implementation of CO_2 storage in the aforementioned formations include, the storage capacity, porosity and permeability of the reservoir rock. In addition, the possibility of CO_2 leakage or defects associated with cement degradation must be examined prior to CO_2 storage in cases of depleted hydrocarbon reservoirs [33]. The possibility of migration of the injected CO_2 and/or brine into the drinking water zones is a major concern that must be taken into account in the case of deep saline aquifers [21].The absence of lateral communication with other mines and/or with the surface (to avoid gas leakage), as well as the minimum depth of the mine top are crucial parameters for CO_2 storage in abandoned coal mines [34]. Regarding the CO_2-mineralisation in rocks, the abundance of Ca–Fe–Mg bearing minerals significantly affects the amount of carbonate minerals produced during reaction of the rock with the injected CO_2 [3].

At the current stage Greece has not developed and adopt significant CO_2 storage sites and policies, while there are only few studies concerning the application of Carbon Capture and Storage (CCS) technologies in this region [35]. In Greece, the regions of Volos (Figure 1) [3], the Mesohellenic Trough [29], the Prinos oil and gas field (Kavala; North Greece) and also parts of Western Greece [36] encompass the appropriate rock types that could serve as potential sites for CO_2 storage. Preliminary estimations of CO_2 storage in Volos basalts indicate storage capacity of ~43,200 tons CO_2 at 300 m depth [3]. Calculations conducted by [37] present high storage capacity of CO_2 in the Pentalofos and Eptachori sedimentary formations of the Mesohellenic Trough at 2 and 3 km depths, respectively. Calculations of CO_2 storage at Klepa-Nafpaktia sandstones in Western Greece indicate storage capacity of 18×10^5 tons of CO_2 at 500 m depth [36]. Additional study indicates that ultramafic rocks from Vourinos ophiolite complex (Western Macedonia; North Greece) could be used for CO_2 sequestration purposes [38].

The present study aims to present, map and investigate specific underground energy storage sites that can be considered to potentially utilise Underground Thermal Energy Storage (UTES), Underground Gas Storage (UGS), Hydrogen and CO_2 underground storage practices in Greece. For this purpose, we considered data derived from petrological and geochemical assessments, field observations and referenced data from extensive literature review, but also through the application of energy storage calculations that take into consideration the physicochemical properties of the potential reservoirs. In this framework, we aim to revise and upgrade preliminary research results provided by [39].

2. Analytical Techniques

Preliminary results of the selective surface rock samples from several parts of the Mesohellenic Trough and the region of Volos have been examined regarding their petrographic properties. Four sandstone samples were collected from the regions between Deskati and Doxiana near Grevena and one sandstone sample was collected from the Lignite Center of Western Macedonia near Ptolemais. Three basaltic rock samples were examined from the regions of Microthives and Porphyrio near Volos. XRD analyses were conducted at CERTH's (Centre for Research and Technology, Hellas) Laboratories, using a Philips X'Pert Panalytical X-ray diffractometer (Malvern Panalytical, Malvern, UK) that operates with Cu radiation at 40 kV, 30 mA, 0.020 step size and 1.0 s step time. Interpretation of XRD results, was accomplished by deploying the DIFFRAC.EVA software v.11 (Bruker, MA, USA), which is based on the ICDD (International Center for Diffraction Data) Powder Diffraction File (2006). Sandstone and basaltic rock samples were examined through petrographic observations in polished thin sections using a Zeiss Axioskop-40 (Zeiss, Oberkochen, Germany), equipped with a Jenoptik ProgRes CF Scan microscope camera at the Laboratories of CERTH. The modal composition of pores was calculated based upon ~1000-point counts on each thin section. Quantification of modal composition was accomplished by using microphotography of SEM images at CERTH's Laboratories with a SEM–EDS (Scanning Electron Microscopy with Energy Dispersive Spectroscopy) JEOL JSM-5600 scanning electron microscope (Jeol, Tokyo, Japan), equipped with an automated energy dispersive analysis system Link Analytical L300 (Oxford Instruments, Abington, UK), with the following operating conditions: 20 kV accelerating voltage, 0.5 nA beam current, 20 s time of measurement, and 5 μm beam diameter. The petrographic description of the sandstone and basaltic rock samples coupled with data provided by extensive literature review are presented below.

3. Geological Background and Petrological Investigation of Rock Types Considered for Storage

3.1. Evaporite Formations

In the Ionian and Pre-Apulian geotectonic zones, evaporites exhibit Triassic and Triassic–Lower Jurassic ages, respectively. The thrust boundary between the Ionian and pre-Apulian zones is marked by intrusion of evaporite occurrences, mostly after the effects of diapirism [40,41]. In the regions of Epirus and Akarnania (Western Greece), Triassic evaporites occur in various localities and in the central part of the Corfu Island [42]. These occurrences provide proof that in the External Hellenides the evaporites represent the lowest detachment level of individual overthrust sheets [43]. Their thickness varies but in most cases is about 2–3 km, although locally it can reach up to 4 km [44–46]. Diapiric forms are usually observed in local scale, resulting in the deformation of the adjacent sedimentary rocks including those of Zavrohon diapir, that has also been confirmed by drilling boreholes [44]. Evaporite formations from Agios Sostis and Laganas regions in Zakynthos island (Western Greece) are represented by the Messinian evaporite unit [47]. These formations consist of nodular and banded evaporate minerals often being in contact with turbidite successions, whereas their formation is associated with present-day active tectonics [44,47].

The principal evaporite mineral phases that have been identified in the Greek evaporite formations are gypsum, halite and anhydrite [40,48]. On their upper stratigraphic levels, dolomite occasionally participates in the form of breccias, which is mostly observed in the Triassic formations. Accessory mineral phases include bassanite and celestite that are not always present. In Northwestern Greece, the porosity of evaporitic formations ranges widely from relatively impermeable to very permeable attributed to local karstification processes [49].

Triassic evaporites from the regions of Anthoussa and Agios Ioannis (West Greece) [50] indicate that these can be distinguished into two main types (Type-23 and Type-24) of Standard Microfacies based on the Flugel classification [51]. Both types are rich in dolomite and solution collapse breccia, whereas calcite, gypsum and anhydrite appear in notable but lower amounts [50]. Type-23 evaporites are laminated by layers of gypsum, anhydrite and dolomite. Dolomite ranges between 40%–60%, evaporite crystals are often calcified and form lenses and rosettes within the mudstone groundmass, whereas relic sparite and evaporite minerals are also observed [50]. Their secondary porosity ranges between 20%–30%. Type-24 evaporites are classified as dolomites and dolomitic limestones. Their depositional texture corresponds to mudstone-grainstone types [50]. Micritic cement is highly but not totally dolomitised, whereas sparitic material usually occurs as cementitious material [50]. Dolomite is present in high amounts reaching 80% [50]. Evaporite minerals reach ~20% and usually form rosette and lenses within the groundmass [50]. Secondary porosity ranges between 15%–40% [50].

Evaporite samples studied from different drilling depths at the region of Kristallopigi (Igoumenitsa; Western Greece) [49] include claystones and siltstones in the upper parts and organic rich evaporites with claystone intercalations in the lower parts. Their mineralogical composition is distinguished into evaporitic and non-evaporitic, including the following assemblages gypsum–anhydriten–bassanite–celestite and quartz–feldspar–calcite–dolomite–magnesite clays, respectively [49]. Clay minerals are classified as kaolinite, illite, smectite and mixed phases [49]. Regarding the evaporitic minerals gypsum is stressed and deformed presenting topotactic relation with bassanite crystals [49]. Bassanite is isolated or forms aggregates, whereas celestite is sub-idiotropic to tabular or prismatic forming aggregates or isolated crystals [49]. Porosity measurements show high variation between 0.31%–52.98% for the studied samples [49].

3.2. *Sedimentary Basins*

The Eocene-Miocene Mesohellenic Trough is an elongated sedimentary basin of 200 km length and 30–40 km width, located in NW Greece [52,53] between the Apulian (non-metamorphic) and the Pelagonian (metamorphic) microcontinental plates [54]. It is a back-arc sedimentary basin evolved during the Upper Oligocene to Miocene period [52,55–57], being superimposed on the Olonos-Pindos external and Pelagonian internal geotectonic units. It is the largest and most important molassic basin formed during the last Alpine orogenic stage of the Hellenides, which occurred between the Mid-Upper Eocene and the Mid-Upper Miocene, extending from Albania in the Northern parts, towards the Thessaly region in Greece at the South. The Mesohellenic Trough comprises of the following four main formations from downwards to surface: Eptachorio, Pentalofo, Tsotillio and Burdigalian [58]. The base of the Eptachorio Formation consists of clastic Upper Eocene to Lower Oligocene sediments (conglomerates, sandstones), as well as base deposits. The upper part of Eptachorio Formation (Figure 2a,b) is located in Taliaros Mountain (Grevena-Kastoria; West Macedonia) and appears in the form of local sedimentary phases comprising of sandstones, marls and limestones. The sequence presents thickness varying from 1000 to 1500 m. The Upper Oligocene to Lower Miocene Pentalofos Formation (Figure 2a,b) includes two types of clastic sedimentary rocks separated by marl–sandstone intercalations. Their thickness ranges from 2250 to 4000 m. The Tsotillio Formation (Lower to Middle Miocene) consists of marls accompanied by conglomerates, sandstones and limestones of variable thickness (200 to 1000 m). The Burdigalian Formation comprises of various phases of sediments such as sandy and silty marls, sand clays, sandstones, conglomerates and limestones [37,52,54]. These sediments are deposited in the form of sedimentary wedges. The aforementioned formations are covered from: (a) Upper Eocene to Middle Miocene alluvial, lacustrine and terrestrial sediments and (b) Pliocene to Lower Pleistocene, fluvial and lacustrine clays, sands and loose conglomerates hosting lignite horizons, which in cases were deposited locally in the Mesohellenic Trough.

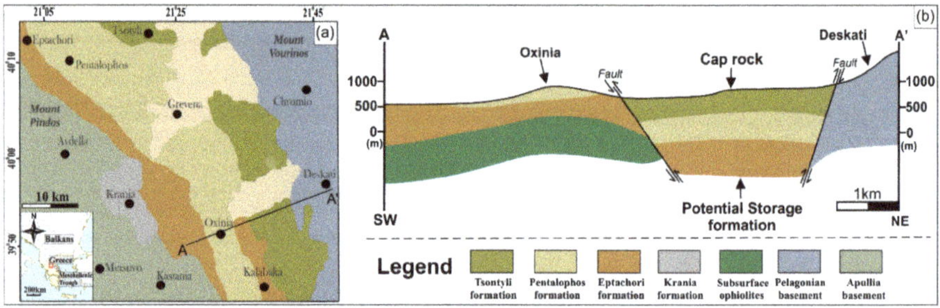

Figure 2. (**a**) Geological map of the Mesohellenic Trough region, WGS'84 and (**b**) Cross section displaying the potential site for energy storage purposes.

The widespread distribution of Mesohellenic Trough sandstones is usually favoured by their high permeability and geochemical characteristics that provide the potential for long-term pH buffer capacity [59]. High volume sandstones are deformed in open anticline structures representing possible porous reservoir rocks [60]. Sedimentary basins of comparable geological features including the Mesohellenic Trough were considered (Table 1) in the current study. In this frame, all of the other sights examined include a suitable geological reservoir consisting of sandstones that are overlain by Tsotyli formation) caprock [61]. Preliminary investigations were conducted in these sights combined with fieldwork and referenced data provided from the literature. These results showed that all the sandstone formations exhibit comparable textural and mineralogical features, presenting restricted variations in terms of porosity.

Their texture ranges from fine grained to medium grained, displaying variable amounts of sub-angular to sub-rounded lithic fragments and straight to suture grain contacts. It also ranges from poor to moderate sorted presenting high amounts of siliceous cement and local patches of calcareous cement material. Their mineralogical assemblage is composed mainly by quartz, alkali-feldspars, calcite (Figure 3), lithic fragments, mica (mostly muscovite and less frequently biotite) and chlorite (Figure 4c,d). Fragments are mostly composed by quartz and feldspar, as well as by clasts of magmatic origin. Quartz appears in the form of monocrystalline angular or polycrystalline sub-angular to sub-round grains. Quartz contacts are straight, suture or interlocking. K-feldspar occurs in the form of variable sized euhedral to subhedral crystals affected from different weathering degrees. Mica and chlorite appear within the sandstone matrix in the form of scarce occurrences developed intergranular between quartz and feldspar crystals.

Table 1. Thermal Capacity properties of Aquifers from selected localities.

Aquifers Considered	Thermal Capacity (MJ)	Productivity (m^3/day)	Heat in Place (MJ)
Sappes	3.370	353	3×10^{-3}
Agioi Theodoroi-Komotini	2.930	1193	3×10^{-3}
Mitrikou lake	3.009	788	9×10^{-4}
Serres	2.435	11,169	1×10^{-3}
W. Thessaloniki Shallow Thermal	2.382	162	4×10^{-4}
W. Thessaloniki-Alexandria	3.577	1516	1×10^{-3}
Mesohellenic Trough_ South Grevena	3.700	57,709	25×10^{-3}
Flysch Botsara syncline	3.822	64,304	18×10^{-3}
Aliveri Evia	2.703	405	1×10^{-3}
Megalopoli	2.291	109	4×10^{-4}
Thimiana Chios	3.021	80	5×10^{-4}
Lesbos-Thermi	2.885	44.5	2×10^{-3}
Kos	2.394	404	1×10^{-3}
Samos	2.461	37.5	7×10^{-4}
Limnos	2.692	38.5	1×10^{-3}
Rhodes	2.249	832	9×10^{-4}
Mesohellenic Trough_Felio	4.175	92,747	18×10^{-3}
W. Thessaloniki_ SG	3.93	11,082	7×10^{-3}
W. Thessaloniki_ DG	4.651	3253	1×10^{-3}
Fili landfill_Attica	2.589	4352	1×10^{-3}
North Mesohellenic basin_ SG	3.700	43,282	22×10^{-3}
North Mesohellenic basin_ DG	4.006	93,778	3×10^{-3}

(Data was collected on potential and possible locations for various underground energy storage technologies in aquifers—see Supplementary Table S1 for more details).

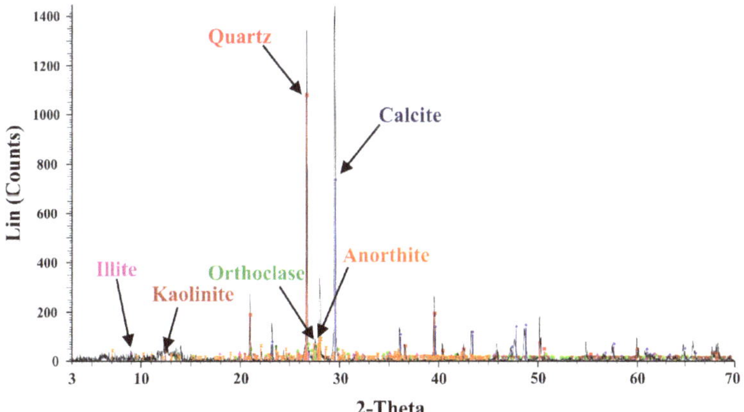

Figure 3. XRD patterns of Mesohellenic Trough sandstone sample (West Macedonia).

3.3. Pleistocene Alkaline Basalt Occurrences

In Greece, these types of rocks are mostly of Triassic age, having been formed at the stage of oceanic rifting. Localities of these basaltic occurrences include Pindos (NW Greece) [62], Koziakas [63], Othris [64], Argolis [65], the South Aegean [66–68], Volos [3] and Evia Island [69]. Alkaline basalt occurrences of relatively recent age (Pleistocene) appear only in the form of scattered outcrops mostly in the Aegean Sea (Central Greece). These occurrences outcrop between North Evoikos and Pagasitikos gulfs (Central Greece) and were formed during Pleistocene by extensional back-arc processes associated with the activity of Northern Anatolia Fault [70–72]. These rocks crop out in the islands of Achilleio, Lichades and Agios Ioannis and consist of massive lavas and pyroclastic rocks [3]. In addition, extensional related basaltic rocks also crop out in the regions of Volos, Kamena Vourla and Psathoura (Central Greece) and they are classified as basaltic trachyandesites and trachyandesites [3]. The aforementioned rock types can serve as potential sites for applications of CO_2 storage, which is further enhanced by their relatively low alteration grade. These volcanic centers are not genetically associated with those developed in the South Aegean arc, which resulted from the subduction of the African plate beneath the Eurasia [3,73,74].

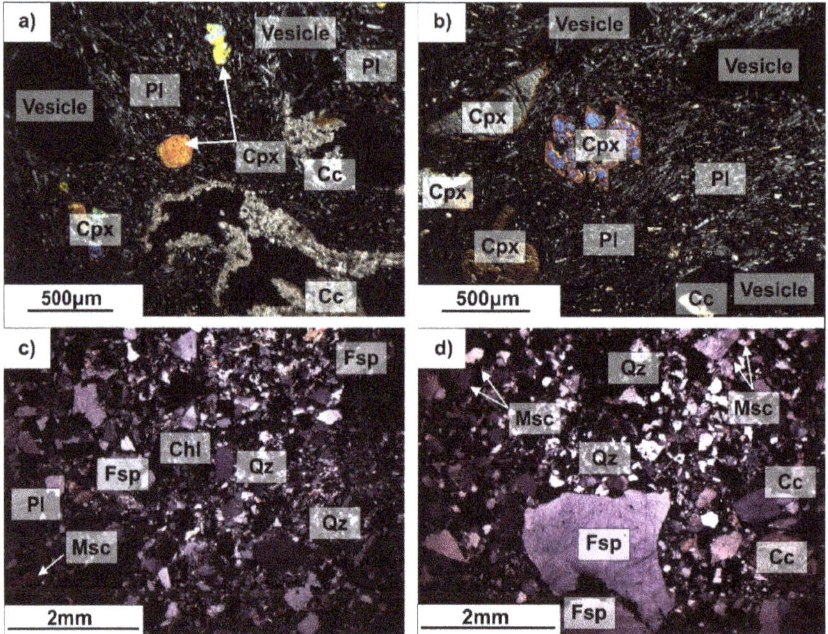

Figure 4. (**a**,**b**) Mesohellenic Trough sandstone photomicrographs including feldspar fragments (Fsp) within quartz (Qz) rich groundmass. Calcite (Cc) and muscovite (Ms) appear in the form of accessory minerals. (**c**,**d**) Basaltic rock samples presenting porphyritic textures composed by plagioclase (Pl) rich groundmass and clinopyroxene (Cpx) phenocrysts. Gas vesicles have been partially filled with secondary calcite (Cc).

Basaltic rocks from the region of Volos present fine grained holocrystalline trachytic or aphanitic groundmass, characterised from porphyritic vesicular textures (Figure 4a,b) Their mineralogial assemblage is composed mainly by subhedral to euhedral clinopyroxene and olivine phenocrystals (Figure 4a,b). Their groundmass is predominantly composed by needle to lath shaped plagioclase and clinopyroxene crystals (~60%). Accessory minerals include orthopyroxene, alkali-feldspar, amphibole, calcite within gas vesicles, pyrite and opaque oxide minerals such as ilmenite and magnetite.

Clinopyroxene is classified mainly as diopside presenting quite high FeO (FeO: 5.37–8.17 wt.%) and variable TiO_2 and Al_2O_3 contents [3]. Olivine is classified as forsterite presenting higher MgO contents compared to those of FeO, whereas Mg# ranges highly [3]. Plagioclase of the basaltic groundmass is CaO rich classified as bytownite or labradorite, whereas the glassy part is characterised by high SiO_2 and Al_2O_3 and total alkali ($Na_2O + K_2O$) contents [3].

The vast majority of basaltic samples present porosity within the range of 15%–23% corresponding to 18% average porosity of the studied suite [3]. Their porosity is highly affected from the occurrence of gas vesicles that present variable size range and are usually filled with secondary calcite.

4. Types of Energy and CO_2 Storage Considered

4.1. Underground Thermal Energy Storage (UTES)

4.1.1. Thermal Capacity Properties of Aquifers

The Aquifer Thermal Energy Storage (ATES) is an open loop system that uses permeable water rock layers as means for storing thermal energy and water. Energy can be transferred into or extracted from aquifers using one or more injection and production wells connected with hydraulic pumps and heat exchangers [75]. This technology also can be combined with Geothermal Heat Pumps (GHP). ATES presents the highest energy efficiency compared to other geothermal technologies but strongly depends on the physical properties of the aquifer (thickness, heat capacity). It is also favourable for large-scale energy storage, such as seasonal storage.

In cases where they are overlain by impermeable sedimentary cap rocks (such as siltstones and shales), aquifers can serve as appropriate formations for thermal storage purposes due to their high porosity. In the current study twenty-two geological formations that present the necessary geological, petrological, mineralogical and textural features have been considered for implementation of underground thermal energy storage (Table 1). Thermal capacity, productivity and heat in place are parameters strongly associated with the pore volume and the physicochemical properties of an aquifer media (Table 1). Thermal energy capacity calculations in the studied aquifers was based on the following equation [76] (see Appendix A Table A1):

$$q = (C_p \cdot \text{rho})_{\text{rock}} \cdot (1 - \varphi) + (C_p \cdot \text{rho})_{\text{water}} \cdot \varphi \quad (1)$$

Based on referenced data specifications [76] a heat capacity value of 0.19 m (mineral matrix) and an average density of 1.41 kg/m^3 were considered as appropriate values for the sandstone aquifers. Results of the current research study indicate that porosity ranges from 8% to 22%, with an average value of 14%. Therefore, the specific thermal capacity per cubic meter for the aquifers considered in the current study is calculated as follows: $q = 2.36 \times 10^{-5}$ MJ/m^3·°C.

The parameters that have been taken into consideration in order to make our energy storage calculations are transmissivity (Table 2), productivity, porosity, aquifer thickness and permeability. In particular, transmissivity is the product of the aquifer thickness (D) and the average value of hydraulic conductivity (K). It is expressed in m^2/day of aquifer thickness [77]. Transmissivity describes an aquifer's capacity to transmit water.

$$T = K \cdot D \quad (2)$$

The productivity of an aquifer depends on its ability to store and transmit water, as well as on the physicochemical characteristics of the geological formation [78]. Productivity is significantly associated with the porosity type, which is classified into primary and secondary [78,79]. In the first case, water is stored within the interstices between grains, whereas secondary porosity refers to the water stored and/or flowing through fractures. Estimation of productivity provides a way to compare the quality of wells from reservoirs with similar properties. Simplified versions of the equations are listed below [80] (see Appendix A Table A1):

$$Q_o = KV1 \cdot Kh \cdot (PF - PS)/VISO \quad (3)$$

$$Q_g = KV2 \cdot Kh \cdot ((PF - PS)^2)/(T + KT2) \quad (4)$$

Heat in place refers to the assessment of the stored heat within a geothermal reservoir. The heat pump potential of geothermal systems can be assessed by estimating the recoverable heat from the total thermal energy stored in a significant volume of porous and permeable reservoir [81–83]. The thermal extraction rate of a specific reservoir is highly associated with its heat transfer properties affected from the fracture network of the rock mass, as well as the flow regime for heat transfer. The energy stored within a geothermal reservoir is calculated by the following equation [84]:

$$E = \rho \cdot c \cdot V \cdot (T_R - T_{ref}) \tag{5}$$

An alternative calculation for the available heat in place (H) expressed in Joule is provided according to the following equation [85]:

$$H = V_{\mathbf{rock}} \cdot P_{\mathbf{rock}} \cdot Cp_{rock} \cdot (Tz - Tr) \tag{6}$$

Heat injection temperature is associated with the amounts of Underground Thermal Energy storage, as well as with the heat and thermal energy recovery [86]. In the current study, a scenario of injecting 70 °C was considered. The decrease rate of the initially achieved temperature was controlled by two major factors. The first factor includes the temperature distribution in association with the ratio distance of the media, whereas the second concerns the heat loss resulted from the thermal conduction [86,87]. The latter strongly depends on the time period and the initial temperature of the aquifer within the sandstone reservoir. Taking these into consideration, a 70 °C temperature scenario was implemented in every case. The average temperature parameters were applied in Equation (6) also considering the local aquifer and injection temperatures. The expected temperature of an aquifer after the injection of 70 °C is estimated to be ~10 °C warmer. In the case of the "Sappes" aquifer the temperature is 33 °C (Supplementary Table S1). Thus, the temperature to be applied in Equation (1) is estimated 43 °C, which equals to 3.370 MJ thermal capacity.

Table 2. Transmissivity properties of the studied aquifers (see Supplementary Table S1 for more properties).

Aquifers Considered	Transmissivity (m^2)
Aquifer Sappes	6.9×10^{-14}
Aquifer Agioi Theodoroi-Komotinis	2.21×10^{-13}
Aquifer Mitrikou lake	1.77×10^{-13}
Aquifer Serres	4.14×10^{-13}
Aquifer Western Thessaloniki_Shallow Thermal	5.92×10^{-14}
Aquifer_Western Thessaloniki_Alexandria	1.18×10^{-13}
Aquifer_Mesohellenic Trough_South Grevena	1.97×10^{-12}
Aquifer Flysch Botsara syncline	2.96×10^{-13}
Aquifer Aliveri Evia	3.45×10^{-13}
Aquifer Megalopoli	7.89×10^{-14}
Aquifer Thimiana Chios	7.4×10^{-14}
Aquifer Lesbos-Thermi island	5.92×10^{-15}
Aquifer Kos island	6.9×10^{-13}
Aquifer Samos island	2.96×10^{-14}
Aquifer Limnos island	2.96×10^{-14}
Aquifer Rhodes	1.18×10^{-13}
Aquifer_Mesohellenic Trough_Felio	1.48×10^{-12}
Aquifer_Western Thessaloniki_SG	2.96×10^{-13}
Aquifer_Western Thessaloniki_DG	8.29×10^{-14}
Aquifer_Fili landfill_Attica	2.24×10^{-13}
North Mesohellenic basin_SG	1.97×10^{-12}
North Mesohellenic basin_DG	1.48×10^{-12}

A storage system can potentially include more than one reservoir rocks characterised from different physicochemical properties. The content of a reservoir rock can be estimated through direct or indirect techniques. This can be accomplished via laboratory measurements on core rock samples [88]. The assessment of an explored rock reservoir strongly depends on its mineralogical and chemical composition, coupled with its morphological and sedimentological properties [89].

The studied aquifers of the current research, present thermal capacity values that range between 2.249 and 4.651 MJ. Aquifers of Mesohellenic Trough, Western Thessaloniki basin, as well as the aquifer of Botsara flysch display the best performance regarding their thermal capacity properties (Figure 5). In particular, Mesohellenic Trough aquifers present thermal capacity contents ranging from 3.7 MJ to 4.175 MJ (Table 1). Similar values correspond to the studied aquifers of North Mesohellenic Trough ranging from 3.7 to 4.006 MJ. The aforementioned results indicate that Mesohellenic Trough aquifers exhibit homogeneity regarding their thermal energy properties throughout their whole geographic distribution. On the other hand, aquifers of Western Thessaloniki basin (Figure 5), exhibit a wide range of thermal capacity contents from 2.382 to 4.651 MJ (Table 1).

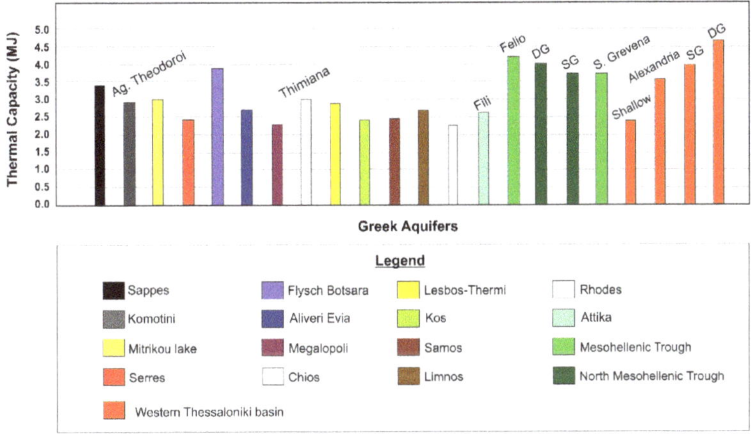

Figure 5. Thermal capacity column chart of the studied Greek aquifers.

4.1.2. UTES Heat Capacity Properties

Underground thermal energy storage (UTES) systems cover a wide range of different technologies including: (a) sensible heat storage that transfers heat to the storage medium or liquid, (b) latent heat storage, where heat is absorbed or released by a phase transition and (c) thermo-chemical storage, in which energy is stored or released through thermochemical reactions.

Assessment of the stored heat is based on numerical-simulation models, providing the potential to address the dynamic changes to the thermal energy as they are affected by recharge and injection processes [90]. The resource volume can change as it shrinks or expands by injection or hot/cold recharge [90]. To quantify these resources it is necessary to determine the amount of the available heat and physicochemical properties of the geothermal rock reservoir [91]. Geothermal resources assessment is based on a volumetric heat content model for porous reservoirs. The heat stored within the rock matrix (index m) and pore water (index w) is calculated by the following equation [92] (see Appendix A Table A1).

$$H = [(1 - P) \cdot \rho_m \cdot C_m + P \cdot \rho_w \cdot C_w] \cdot (T_t - T_o) \cdot A \cdot \Delta z \quad (7)$$

The Area Heat Energy Capacity is calculated by dividing the amount of the Total Heat Energy Capacity with the surface area according to the following equation:

$$\text{Area Heat Energy Capacity}\left(\text{MWh}_{[\text{th}]}/\text{km}^2\right) = \frac{\text{Total Heat Energy Capacity}\left(\text{MWh}_{[\text{th}]}\right)}{\text{Total Area}\left(\text{km}^2\right)} \quad (8)$$

Research results of the current study indicate that the Serres, Komotini and Sappers aquifers exhibit the highest average values of total heat energy (Figure 6). In particular, the Serrese aquifer exhibits the highest potential of 6,052,932 MJ, followed by the Komitini (Agioi Theodoroi) and Sappes aquifers with 4,076,831 and 2,172,617 MJ, respectively (Table 3). Comparison between aquifers and abandoned mines in the studied regions indicates that in general aquifers present significantly higher total heat energy capacity compared to that of the abandoned mines. The abandoned coal mine of Aliver (Evia Island; Central Greece) shows the highest potential for underground heat storage (723,255 MJ) compared to that of Chaidari and Mandra in Attica (Central Greece).

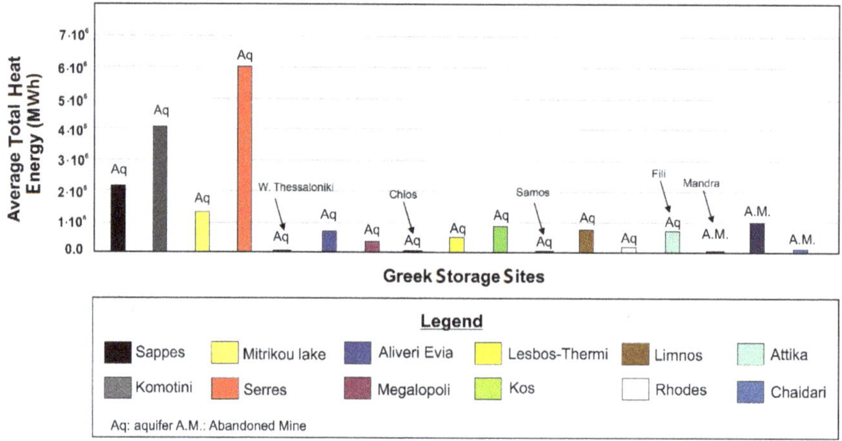

Figure 6. Average total heat energy column chart of the studied Greek aquifers and abandoned mines.

Considering the area average heat values of the studied regions (Table 3), the aquifers tend to exhibit lower values compared to those of abandoned mines, which are significantly higher (Figure 7). In particular, the aquifer average area heat energy ranges from 227,864 to 1,365,479 MJ. The highest values in the studied aquifers correspond to those of Sappes and Komotini (Agioi Theodoroi) regions. Regarding the abandoned mines their average area heat energy ranges from 1,181,993 to 5,440,280 MJ, with the Aliveri abandoned coal mine presenting the highest potential.

The aforementioned results (Table 3) indicate that Aquifer Thermal Energy Storage is one of the most promising technological options that provides large storage capacities [93,94]. By utilising the subsurface space in these sites, Underground Thermal Energy Storage systems potentially provide sustainable heating and cooling energy for different building types. Thus the ATES integration in a district or urban level can be more efficient compared to a conventional separate heating and cooling generation technology [94,95].

The East Macedonia-Thrace region consumes 3765 kWh per household for electricity needs [96]. In Greece 40% of the total energy consumption corresponds to electricity needs, whereas the 60% corresponds to space heating [97]. Thus, in East Macedonia-Thrace the thermal energy consumption is 5647.5 kWh. Taking these into consideration, it is suggested that the Sappes, Serres and Komotini (Agioi Theodoroi) aquifers (Table 3) could cover the space heating energy consumption of this region.

Table 3. Underground Thermal energy storage systems (UTES) Capacity in selected sites (see Supplementary Table S2 for additional details).

Sites Considered (* Aquifer, ** Abandoned Mine)	Avgerage Total Heat Energy Storage Capacity ($MWh_{[th]}$)	Average Area Heat Energy Storage Capacity ($MWh_{[th]}$)
Sappes *	2,172,617	1,284,404
Agioi Theodoroi Komotinis *	4,076,831	1,365,479
Mitrikou lake *	1,335,808	448,048
Serres *	6,052,932	793,685
W. Thessaloniki Shallow Thermal *	13,232	249,200
Aliveri Evia *	723,255	691,274
Megalopoli *	384,557	227,864
Thimiana Chios *	9599	256,027
Lesbos-Thermi *	509,167	1,024,109
Kos *	847,815	836,356
Samos *	21,801	384,041
Limnos *	767,700	473,650
Rhodes *	162,278	477,918
Fili landfill_Attica *	673,119	621,720
Mandra Attica **	34,543	2,997,653
Aliveri **	936,938	5,440,280
Chaidari **	65,198	1,181,993

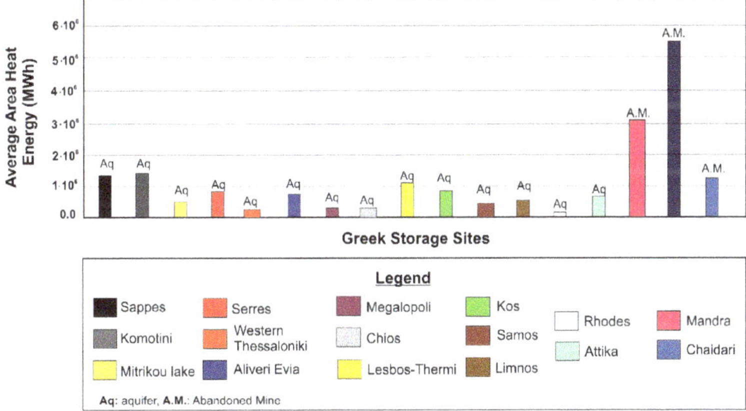

Figure 7. Average area heat energy column chart of the studied Greek aquifers and abandoned mines.

4.2. Underground Gas Storage Potential

Underground gas storage is currently regarded as a mature and widely implemented technology in a global scale. The choice of the appropriate gas reservoir strongly depends on many factors such as [98]: (a) porosity, permeability and storage capacity of the host rock, (b) cap rock occurrence, (c) depth of capture and (d) the physicochemical properties of the underground water.

There are three types of underground natural gas storage at the current stage and these include depleted oil and gas reservoirs, salt-caves and aquifers. Natural gas is injected into the storage regions during the slack season of gas usage (mainly summer) and is produced during the busy season (mainly winter) from storage reservoirs [99]. In depleted natural gas and oil reservoirs conversion of gas/oil field (from production to storage stage) benefits from the existing infrastructure (wells, gathering systems, and pipeline connections), as well as on the massive societal acceptance. These are the most commonly used underground storage types due to their wide availability [100]. The amount of the stored natural gas strongly depends on the formation porosity and permeability, which affect the natural gas flow rate. The flow rate subsequently affects the rate of injection and withdrawal of working gas [101].

Predictions suggest an increasing energy demand within the following years. Therefore, planned infrastructure projects, in which Greece involves as a transit country, will be developed. These projects include the Trans Adriatic Pipeline (TAP) [102–104] the IGB (Interconnector Greece Bulgaria) gas interconnector between Greece-Bulgaria [103,104] and the EastMed gas pipelines [103] to improve security and diversity of Europe's energy.

Results of the current study (Table 4) indicate that potential sites for implementation of underground gas storage technologies in Greece include gas fields, aquifers and evaporites. In particular, the Prinos and South Kavala gas fields present significantly higher storage capacities compared to the other types presented (aquifers and evaporites; Figure 8). These results further confirm the high storage potential of South Kavala gas field for natural gas storage purposes. The South Kavala Gas field (North Greece) is a promising case for underground gas storage in Greece (Table 4). It is a site with proven feasibility and is expected to be deployed in the following years. The available data show that the cumulative gas production to date is approximately 847 Mm^3 (million m^3) (Recovery Factor RF 85%) and the estimated remaining gas volume is 148 Mm^3, respectively [105]. Evaporite formations are the second most important types presenting storage capacity values ranging from 649,668 to 757,327 $MWh_{(e)}$. Regarding the studied aquifers, those included in the sedimentary formations of Western Thessaloniki basin and Mesohellenic Trough exhibit the highest potential for underground gas storage.

Taking into consideration that the electricity consumption per household is 3765 kWh for East Macedonia and Thrace [96], it is concluded that the Prinos and S. Kavala gas fields (Table 4) could cover the electricity needs for the households in this region.

Table 4. Underground gas storage (UGS) Capacity in selected sites (see Supplementary Table S3 for additional details).

Sites Considered (* Gas Field, ** Aquifer, *** Salt Structure)	Total Gas Volume (Mm^3)	Working Gas Volume (Mm^3)	Cushion Gas Volume (Mm^3)	Energy Storage Capacity ($MWh_{[e]}$)
South Kavala *	847	720	127	2,672,920
Epanomi *	500	250	250	928,097
Katakolo *	300	150	150	556,858
Prinos *	2280	1300	980	4,826,105
Western Thessaloniki_ Alexandria **	3	1	2	3712
Mesohellenic Trough_ South Grevena **	44	13	31	48,261
Flysch Botsara syncline **	33	10	23	37,123
Mazarakia ***	281	175	106	649,668
Heraklion ***	328	204	124	757,327
Mesohellenic Trough Filio **	25	8	17	29,699
Western Thessaloniki SG **	136	41	95	152,208
Western Thessaloniki DG **	29	93	20	345,252
North Mesohellenic basin SG **	130	39	91	144,783
North Mesohellenic basin DG **	214	64	150	237,593
Delvinaki ***	218	175	106	649,668

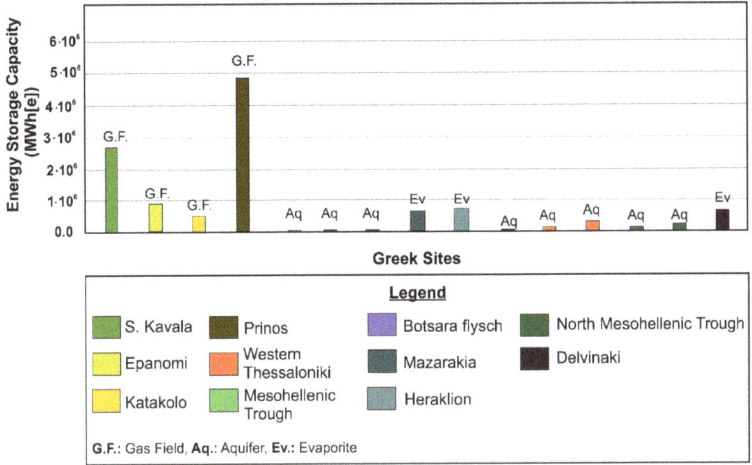

Figure 8. Average energy storage capacity ($MWh_{[e]}$) column chart of the studied Greek aquifers, evaporites and abandoned mines.

4.3. Hydrogen Storage Potential

According to the geological data considered, the potential sites for the development of underground hydrogen storage practices are those of Trifos, Achira (Aitoloakarnania Central-Western Greece), as well as the Kefalonia and Corfu islands. These locations have been already deployed for on-shore wind energy production, whose excess energy can be used to produce hydrogen within an environmentally friendly framework. In order to modelise the potential of storing hydrogen within the aforementioned storage sites a capsule-shaped cavern is assumed as a case scenario. This particular shape is proposed because research studies have shown that these are more stable in terms of stress risks compared to other shapes considered for cavern development [106]. A capsule cavern of elliptical shape was assumed [106]. The cavern volume is 550,000 m^3, with diameter of 60 and 300 m height, at a depth of 650 m. The average temperature of hydrogen stored in such evaporite cavern is calculated using Equation (9) [17]:

$$T_{average} = 288 + 0.025 \times (\text{depth} - \text{cavern height}/2) \tag{9}$$

which is calculated at 27 °C, taking into consideration the geothermal gradient. In this equation, $T_{average}$ stands for average gas temperature of 15 °C, depth in meters and cavern height in meters.

Based on reference data considerations [17,106], an amount of 367,000 m^3 as the average working hydrogen volume and an amount of 183,000 m^3 for cushion volume were estimated. The working volume refers to the maximum usable storage capacity expressed in energy content. The storage capacity can also be expressed in energy terms and significantly by calculating it in $MWh_{(e)}$. Our case scenario presents comparable features with an already existing hydrogen storage site in Chevron, Texas (USA), which enables the storage of 580,000 m^3 hydrogen volume, at 83,300 $MWh_{(th)}$ energy storage capacity and ~850–1150 m depth [106]. Based on this site specifications it is estimated that our case study presents the potential for 80,000 $MWh_{(th)}$ energy storage capacity for 650 m depth and 550,000 m^3 hydrogen volume Considering that 1 $MWh_{(e)}$ = 1/3 $MW_{(th)}$ [107], the aforementioned value corresponds to 26,600 $MWh_{(e)}$. This scenario applies for all the above-mentioned sites (Trifos, Achira, Kefalonia and Corfu) that could potentially be deployed.

Taking into consideration that the average electricity consumption of Ionian islands is 3929 kWh per household [96] and the islands of Kefalonia and Corfu can store up to 53,200 $MWh_{(e)}$, it is concluded that hydrogen could cover the electricity needs of 6770 households.

4.4. Recommended CO_2 Storage Sites

Geological CO_2 storage within permeable rock types such as basalts (CarbFix project, Iceland) [27,28] and sandstones [29,30,108] is one of the most efficient technologies for mitigating anthropogenic CO_2 emissions. Enhanced oil recovery provides a crucial option for secure CO_2 storage in oil fields. In this framework several high impact EOR (Enhanced Oil Recovery) projects have been implemented [26,109]. Specific rock types (basalts, sandstones and evaporites) from many Greek regions were considered in the current study regarding their potential for CO_2 storage.

There is a detailed literature regarding the potential of selected Greek regions. Pleistocenic alkaline basaltic outcrops in the region of Volos is considered as a promising site to apply carbon storage [3]. Their abundance in Ca-bearing minerals, low alteration grade and high porosity provide the appropriate physicochemical properties for the implementation of CO_2 storage via mineralisation. Preliminary calculations indicate a storage potential of 82,800 and 27,600 tons of CO_2 maximum mass [3]. The aforementioned indications for implementation of carbon storage are further supported by the indications of enhanced heat based on calculations of groundwater temperatures from irrigation wells. Similar elevated temperatures have been calculated in adjacent regions of Kamena Vourla (Central Greece; East Thessaly) and Lichades Islands (Central Greece; North Evoikos Gulf) [110].

Based on geochemical simulations conducted by [29], presenting reactions within the sandstone brine system of the Mesohellenic Trough it is concluded that the Pentalofos and Tsotyli sedimentary formations could serve as potential sites for long term CO_2 storage produced by the adjacent power plants [29]. The potential of CO_2 storage within sandstone formations is highly affected from the abundance of feldspars and plagioclase crystals [36,111]. Research studies indicate that K-feldspars react with the injected CO_2 producing clay minerals (kaolinite, illite), $NaAlCO_3(OH)_2$ and K_2CO_3 [36]. Kaolinite precipitation is further enhanced from plagioclase dissolution which also produces considerable amounts of calcite [36]. Further estimations on CO_2 capacity in the Mesohellenic Trough indicate that Pentalofos formation is a geological reservoir formation capped by Tsotylli formation which serves as a cap-rock [37].

Additional sites with potential of CO_2 storage in Greece include the Miocene sandstones of the Prinos and South Kavala regions, at ~1600 m depth with 35 Mt CO_2 storage capacity [19]. The offshore-Prinos basin can host 30 Mt CO_2 within oil reservoirs, as well as 1350 Mt CO_2 in deep saline Miocene aquifers at 2400 m depth [19]. The Epanomi gas field (Figure 1) presents 2 Mt CO_2 capacity within limestones at 2600 m depth, whereas the deep saline aquifers within sedimentary formations of Thessaloniki basin present 640 Mt CO_2 storage capacity at 900–2400 m depth [19]. In South Greece the Katakolo oil and gas field could serve as potential site for CO_2 storage purposes within the Cretaceous-Eocene limestones presenting 3.2 Mt CO_2 capacity [19].

Based on the aforementioned literature data, in combination with additional key parameters guidelines are proposed suggesting which perspective sites are most suitable for CO_2 storage: (a) Mesohellenic Trough sandstones, (b) Volos basalts, (c) Western Thessaloniki saline aquifers and (d) depleted gas field of Prinos. These parameters include the geographical distribution of the regions, the CO_2 storage techniques that can be implemented, as well as the possibility to develop cost-effective scenarios.

The Volos basalts, the Mesohellenic Trough and the Western Thessaloniki basin are located close to industrial regions, indicating that CO_2 storage can be favoured by reduced costs of transport. These industrial regions include the cities of (a) Volos (10 km distance from the corresponding basalt occurrences), (b) Grevena, Ptolemais, Kozani (which lie close to the Mesohellenic Trough sandstones) and (c) Thessaloniki (which is located at the geographical boundaries of Western Thessaloniki basin). The implementation of CO_2 storage within Mesohellenic Trough sandstones through combination of geological storage (within the permeable Tsotilli and Pentalofos formations), or carbon mineralisation, provides significant advantages compared to the other regions.

Concerning the case of the depleted oil reservoir in the off-shore Prinos basin, the relatively high storage capacity coupled with the use of existing infrastructures, suggests that storage of CO_2 can be implemented in a cost-effective framework. In addition, the capability of such reservoirs to host hydrocarbons deposits for long time periods minimises risks for safety hazards associated with CO_2 containment compared to other sites [112]. Deep saline aquifers often include water of unsuitable quality for agricultural and drinking water usage [21]. Thus, the aquifers of Western Thessaloniki basin present significant advantage to be exploited for CO_2 storage.

5. Conclusions

This study presents, maps and investigates several Greek regions regarding their capacity on underground thermal energy storage (UTES), underground gas storage (UGS), hydrogen storage and CO_2 storage. To achieve that preliminary results and petrographical data were presented and compared with literature data. The main concluding remarks are:

- Thermal energy capacity was examined in 22 sandstone aquifers throughout several regions of Northern, Central Greece and Aegean islands. Research results indicate that the aquifers of Mesohellenic Trough, Western Thessaloniki basin and those of Botsara flysch exhibit the highest thermal capacity values, which can reach up to 4175 MJ.
- Heat capacity was investigated in 14 aquifers throughout the Northern and Southern Greece, as well as on three abandoned mines from Attica region and Aliveri regions (Central Greece). In general, the aquifers tend to present higher average total heat energy values that reach up to ~6.05×10^6 $MWh_{(th)}$ compared to those of the abandoned mines. In particular, the Sappes, Serres and Komotini (Agioi Theodoroi) aquifers could cover the space heating energy consumption of East Macedonia-Thrace region.
- Underground gas storage technologies present better performance in the gas fields of South Kavala and Prinos basins compare to the studied evaporite formations and aquifers. These two gas fields could cover the electricity needs of the households in the region of East Macedonia and Thrace.
- Hydrogen storage capacity was based on a capsule-shaped cavern scenario in Trifos and Achira regions (Central-Western Greece), as well as in Kefalonia and Corfu islands. Calculations indicate storage capacity of 26,600 $MWh_{(e)}$ for each of the studied regions. These calculated hydrogen capacity values for Corfu and Kefalonia islands, could efficiently cover the electricity needs of 6770 households in the Ionian Sea region.
- Petrological studies of the Mesohellenic trough sandstones coupled with basaltic rocks from Volos region (Central Greece) indicate that these rocks could serve as potential sites for CO_2 storage via CO_2-mineralisation. This is attributed to their high porosity, low alteration grade and abundance on Fe–Mg–Ca–K silicate minerals.
- Based on CO_2 capacity data provided by the literature, geographical distribution criteria, potential for implementation of CO_2 storage techniques and cost-considerations, we recommend the: (a) Volos basalts, (b) Mesohellenic trough sandstones, (c) saline aquifers of Western Thessaloniki basin and (d) the oil reservoir of Prinos basin, as the most promising sites for CO_2 storage in Greece.

Supplementary Materials: The following are available online at http://www.mdpi.com/1996-1073/13/11/2707/s1, Table S1: Prop (Properties), Table S2: Assess_UGS, Table S3: Assess_H2, Table S4: Assess_Therm (Thermal Capacity), Table S5: Coordinates.

Author Contributions: Conceptualization, A.A. and P.K.; Investigation, A.A., P.K., N.K., P.T., D.K. and C.K.; Methodology, A.A., P.K., D.K. and C.K.; Supervision, A.A., P.K. and N.K.; Validation, D.K. and P.P.; Writing—original draft, A.A., P.K., N.K., P.T., D.K., C.K. and P.P. All authors have read and agreed to the published version of the manuscript.

Funding: This research received no external funding.

Acknowledgments: We would like to express our sincerest thanks to the Reviewers and the Editor for their constructive comments that substantially helped to improve the current paper. The preliminary research results provided in this paper were published in the frame of the EU Horizon 2020-funded program ESTMAP. Many thanks are also given to Georgia Kastanioti (Geologist) and Pavlos Krassakis (Geologist) for their participation of the previously mentioned EU Horizon 2020-funded project.

Conflicts of Interest: The authors declare no conflict of interest.

Appendix A

Table A1. Symbology, description of symbols and units used in equations presented in this study.

Symbol	Description	Unit
Cp	heat capacity	MJ/kg·°C
rho	dry bulk density	kg/m^3
phi	porosity	%
T	transmissivity	m^2/day
K	hydraulic conductivity	m/day
D	aquifer thickness	m
Kh	permeability-thickness product	m^2-m
PFa	average formation pressure	Pa
T	temperature	°C
VISO	oil viscosity	mm^2/s
Qg	productivity gas	m^3/day
c	volumetric specific heat of the reservoir rock	MJ/kg·°C
V	volume of the reservoir	m^3
TR	the characteristic reservoir temperature	°C
TREF	the reference temperature	°C
H	heat in place	MJ
Vrock	the denoting rock volume	m^3
Prock	the rock density	kg/m^3
H	total heat energy storage capacity	MJ
P	effective porosity	%
ρm	density of the rock matrix	kg/m^3
ρw	density of water	kg/m^3
cm	specific heat capacity of the rock matrix	MJ/kg·°C
cw	specific heat capacity of water	MJ/kg·°C
Tt	mean temperature of the compartment	°C
A	surface area under consideration	m^2
Δz	aquifer thickness	m
Taverage	average temperature of stored hydrogen	°C
depth	depth	m
cavernHeight	height of the cavern	m
	KV1 = 7.5×10^{-6} m^3/day	
	KV2 = 1×10^{-5} m^3/day	
	Density of water: 997 kg/m^3	
	Density of rock matrix (e.g., sandstone): 2323 kg/m^3	

References

1. Davis, W. The Relationship between Atmospheric Carbon Dioxide Concentration and Global Temperature for the Last 425 Million Years. *Climate* **2017**, *5*, 76. [CrossRef]
2. IPCC. *Climate Change 2013: The Physical Science Basis. Contribution of Working Group I to the Fifth Assessment Report of the Intergovernmental Panel on Climate Change*; Cambridge University Press: New York, NY, USA, 2013; p. 1535.
3. Koukouzas, N.; Koutsovitis, P.; Tyrologou, P.; Karkalis, C.; Arvanitis, A. Potential for Mineral Carbonation of CO_2 in Pleistocene Basaltic Rocks in Volos Region (Central Greece). *Minerals* **2019**, *9*, 627. [CrossRef]
4. Rosenbauer, R.J.; Thomas, B.; Bischoff, J.L.; Palandri, J. Carbon sequestration via reaction with basaltic rocks: Geochemical modeling and experimental results. *Geochim. Cosmochim. Acta* **2012**, *89*, 116–133. [CrossRef]

5. Bott, C.; Dressel, I.; Bayer, P. State-of-technology review of water-based closed seasonal thermal energy storage systems. *Renew. Sustain. Energy Rev.* **2019**, *113*, 109241. [CrossRef]
6. Montcoudiol, N.; Burnside, N.M.; Győre, D.; Mariita, N.; Mutia, T.; Boyce, A. Surface and groundwater hydrochemistry of the Menengai Caldera Geothermal Field and surrounding Nakuru County, Kenya. *Energies* **2019**, *12*, 3131. [CrossRef]
7. Xu, X.; Wei, Z.; Ji, Q.; Wang, C.; Gao, G. Global renewable energy development: Influencing factors, trend predictions and countermeasures. *Resour. Policy* **2019**, *63*, 101470. [CrossRef]
8. Nataraj Barath, J.G.; Husev, O.; Manonmani, N. Overview of Energy Storage Technologies For Renewable Energy. *IJISET* **2015**, *2*, 6.
9. Mofijur, M.; Mahlia, T.M.I.; Silitonga, A.S.; Ong, H.C.; Silakhori, M.; Hasan, M.H.; Putra, N.; Rahman, S.A. Phase Change Materials (PCM) for Solar Energy Usages and Storage: An Overview. *Energies* **2019**, *12*, 3167. [CrossRef]
10. Dincer, I.; Rosen, M. *Thermal Energy Storage: Systems and Applications*; John Wiley & Sons: Hoboken, NJ, USA, 2010.
11. Paksoy, H.Ö.; Beyhan, B. Thermal energy storage (TES) systems for greenhouse technology. In *Advances in Thermal Energy Storage Systems. Methods and Applications*; Cabeza, L.F., Ed.; Woodhead Publishing, Elsevier: Cambridge, UK, 2015; p. 612.
12. De Schepper, G.; Paulus, C.; Bolly, P.-Y.; Hermans, T.; Lesparre, N.; Robert, T. Assessment of short-term aquifer thermal energy storage for demand-side management perspectives: Experimental and numerical developments. *Appl. Energy* **2019**, *242*, 534–546. [CrossRef]
13. Park, J.-W.; Rutqvist, J.; Ryu, D.; Park, E.-S.; Synn, J.-H. Coupled thermal-hydrological-mechanical behavior of rock mass surrounding a high-temperature thermal energy storage cavern at shallow depth. *Int. J. Rock Mech. Min. Sci.* **2016**, *83*, 149–161. [CrossRef]
14. Arvanitis, A. Investigation of the Possibilities for the Subsurface Energy Storage in the frame of the EC-funded ESTMAP Project. In Proceedings of the 11th National Conference on Renewable Energy Sources, Thessaloniki, Greece, 14–16 March 2018.
15. Breeze, P. Power System Energy Storage Technologies. In *Power Generation Technologies*, 3rd ed.; Breeze, P., Ed.; Elsevier Ltd.: Amsterdam, The Netherlands, 2019; pp. 219–249.
16. Lemieux, A.; Sharp, K.; Shkarupin, A. Preliminary assessment of underground hydrogen storage sites in Ontario, Canada. *Int. J. Hydrog. Energy* **2019**, *44*, 15193–15204. [CrossRef]
17. Caglayan, D.; Weber, N.; Heinrichs, H.; Linssen, J.; Robinius, M.; Kukla, P.; Stolten, D. Technical potential of salt caverns for hydrogen storage in Europe. *Int. J. Hydrog. Energy* **2020**, *45*, 6793–6805. [CrossRef]
18. Crotogino, F.; Donadei, S.; Bünger, U.; Landinger, H. Large-Scale Hydrogen Underground Storage for Securing Future Energy Supplies. In Proceedings of the 18th World Hydrogen Energy Conference, Essen, Germany, 16–21 May 2010.
19. Arvanitis, A.; Koukouzas, N.; Koutsovitis, P.; Karapanos, D.; Manoukian, E. Combined CO_2 Geological Storage and Geothermal Energy Utilization in Greece. In Proceedings of the 15th International Congress of the Geological Society of Greece Harokopio University of Athens, Athens, Greece, 22–24 May 2019.
20. Hasan, M.M.F.; Boukouvala, F.; First, E.L.; Floudas, C.A. Nationwide, Regional, and Statewide CO_2 Capture, Utilization, and Sequestration Supply Chain Network Optimization. *Ind. Eng. Chem. Res* **2014**, *53*, 7489–7506. [CrossRef]
21. Celia, M.A.; Bachu, S.; Nordbotten, J.M.; Bandilla, K.W. Status of CO_2 storage in deep saline aquifers with emphasis on modeling approaches and practical simulations. *Water Resour. Res.* **2015**, *51*, 6846–6892. [CrossRef]
22. Jalili, P.; Saydam, S.; Cinar, Y. CO_2 Storage in Abandoned Coal Mines. In Proceedings of the Underground Coal Operators' Conference Wollongong, Wollongong, Australia, 10–11 February 2011; p. 410.
23. Xie, L.Z.; Zhou, H.; Xie, H. Research advance of CO_2 storage in rock salt caverns. *Yantu Lixue/Rock Soil Mech.* **2009**, *30*, 7.
24. Baran, P.; Zarębska, K.; Krzystolik, P.; Hadro, J.; Nunn, A. CO_2-ECBM and CO_2 sequestration in Polish coal seam—Experimental study. *J. Sustain. Min.* **2014**, *13*, 22–29. [CrossRef]
25. Raza, A.; Gholami, R.; Rezaee, R.; Bing, C.H.; Nagarajan, R.; Hamid, M.A. Well selection in depleted oil and gas fields for a safe CO_2 storage practice: A case study from Malaysia. *Petroleum* **2017**, *3*, 167–177. [CrossRef]

26. Györe, D.; Stuart, F.M.; Gilfillan, S.M.V.; Waldron, S. Tracing injected CO_2 in the Cranfield enhanced oil recovery field (MS, USA) using He, Ne and Ar isotopes. *Int. J. Greenh. Gas Control* **2015**, *42*, 554–561. [CrossRef]
27. Snæbjörnsdóttir, S.Ó.; Wiese, F.; Fridriksson, T.; Ármansson, H.; Einarsson, G.M.; Gislason, S.R. CO_2 storage potential of basaltic rocks in Iceland and the oceanic ridges. *Energy Procedia* **2014**, *63*, 4585–4600. [CrossRef]
28. Ragnheidardottir, E.; Sigurdardottir, H.; Kristjansdottir, H.; Harvey, W. Opportunities and challenges for CarbFix: An evaluation of capacities and costs for the pilot scale mineralization sequestration project at Hellisheidi, Iceland and beyond. *Int. J. Greenh. Gas Control* **2011**, *5*, 1065–1072. [CrossRef]
29. Koukouzas, N.; Kypritidou, Z.; Purser, G.; Rochelle, C.A.; Vasilatos, C.; Tsoukalas, N. Assessment of the impact of CO_2 storage in sandstone formations by experimental studies and geochemical modeling: The case of the Mesohellenic Trough, NW Greece. *Int. J. Greenh. Gas Control* **2018**, *71*, 116–132. [CrossRef]
30. Garcia-Rios, M.; Luquot, L.; Soler, J.; Cama, J. Laboratory-Scale Interaction between CO_2-Rich Brine and Reservoir Rocks (Limestone and Sandstone). *Procedia Earth Planet. Sci.* **2013**, *7*, 109–112. [CrossRef]
31. Boschi, C.; Dini, A.; Dallai, L.; Ruggieri, G.; Gianelli, G. Enhanced CO_2-mineral sequestration by cyclic hydraulic fracturing and Si-rich fluid infiltration into serpentinites at Malentrata (Tuscany, Italy). *Chem. Geol.* **2009**, *265*, 209–226. [CrossRef]
32. Dichicco, M.C.; Laurita, S.; Paternoster, M.; Rizzo, G.; Sinisi, R.; Mongelli, G. Serpentinite Carbonation for CO_2 Sequestration in the Southern Apennines: Preliminary Study. *Energy Procedia* **2015**, *76*, 477–486. [CrossRef]
33. Bennaceur, K. CO_2 Capture and Sequestration. In *Future Energy*, 2nd ed.; Letcher, T.M., Ed.; Elsevier: Amsterdam, The Netherlands, 2014; pp. 583–611.
34. Piessens, K.; Dusar, M. CO_2-sequestration in abandoned coal mines. In Proceedings of the International Coalbed Methane Symposium, Tuscaloosa, AL, USA, 5–8 May 2003; p. 11.
35. Kelektsoglou, K. Carbon Capture and Storage: A Review of Mineral Storage of CO_2 in Greece. *Sustainability* **2018**, *10*, 4400. [CrossRef]
36. Petrounias, P.; Giannakopoulou, P.; Rogkala, A.; Kalpogiannaki, M.; Koutsovitis, P.; Damoulianou, M.-E.; Koukouzas, N. Petrographic Characteristics of Sandstones as a Basis to Evaluate Their Suitability in Construction and Energy Storage Applications. A Case Study from Klepa Nafpaktias (Central Western Greece). *Energies* **2020**, *13*, 1119. [CrossRef]
37. Tasianas, A.; Koukouzas, N. CO_2 storage capacity estimate in the lithology of the Mesohellenic Trough, Greece. *Energy Procedia* **2016**, *86*, 334–341. [CrossRef]
38. Koukouzas, N.; Gemeni, V.; Ziock, H.J. Sequestration of CO_2 in magnesium silicates, in Western Macedonia, Greece. *Int. J. Miner. Process.* **2009**, *93*, 179–186. [CrossRef]
39. ESTMAP, Energy Storage Mapping and Planning Home Page. Available online: http://estmap.eu/deliverables.html (accessed on 5 May 2020).
40. Karakitsios, V. Western Greece and Ionian Sea petroleum systems. *AAPG Bull.* **2013**, *97*, 1567–1595. [CrossRef]
41. Karakitsios, V.; Rigakis, N. Evolution and petroleum potential of Western Greece. *J. Pet. Geol.* **2007**, *30*, 197–218. [CrossRef]
42. Tserolas, P.; Mpotziolis, C.; Maravelis, A.; Zelilidis, A. Preliminary Geochemical and Sedimentological Analysis in NW Corfu: The Miocene Sediments in Agios Georgios Pagon. In Proceedings of the 14th International Congress of Geological Society of Greece, Thessaloniki, Greece, 25–27 May 2016; pp. 402–411.
43. Karakitsios, V.; Roveri, M.; Lugli, S.; Manzi, V.; Gennari, R.; Antonarakou, A.; Triantaphyllou, M.; Agiadi, K.; Kontakiotis, G.; Kafousia, N.; et al. A Record of the Messinian Salinity Crisis in the Eastern Ionian Tectonically Active Domain (Greece, Eastern Mediterrenean). *Basin Res.* **2017**, *29*, 203–233. [CrossRef]
44. Kokinou, E.; Kamberis, E.; Kotsi, F.; Lioni, K.; Velaj, T. The Impact of Evaporites in the Greek and Albanian Oil Systems. In Proceedings of the 79th EAGE Conference and Exhibition 2017, Paris, France, 12–15 June 2017.
45. Velaj, T. The structural style and hydrocarbon exploration of the subthrust in the Berati Anticlinal Belt, Albania. *J. Pet. Explor. Prod. Technol.* **2015**, *5*, 123–145. [CrossRef]
46. Velaj, T. Tectonic style in Western Albania Thrustbelt and its implication on hydrocarbon exploration. *AAPG Search Discov. Artic.* **2011**, 123–148.
47. Kontopoulos, N.; Zelilidis, A.; Piper, D.J.W.; Mudie, P.J. Messinian evaporites in Zakynthos, Greece. *Palaeogeogr. Palaeoclimatol. Palaeoecol.* **1997**, *129*, 361–367. [CrossRef]

48. Bourli, N.; Kokkaliari, M.; Iliopoulos, I.; Pe-Piper, G.; Piper, D.J.W.; Maravelis, A.G.; Zelilidis, A. Mineralogy of siliceous concretions, cretaceous of ionian zone, western Greece: Implication for diagenesis and porosity. *Mar. Pet. Geol.* **2019**, *105*, 45–63. [CrossRef]
49. Tsipoura-Vlachou, M. Diagenesis of the Marly-Gypsum Formations, Igoumenitsa Area, N.W. Greece. *Bull. Geol. Soc. Greece* **2007**, *40*, 1009–1021. [CrossRef]
50. Getsos, K.; Papaioannou, F.; Zelilidis, A. Triassic Carbonate and Evaporite Sedimentation in the Ionian Zone (Western Greece): Palaeogeographic and Palaeoclimatic Implication. *Bull. Geol. Soc. Greece* **2004**, *36*, 699–707. [CrossRef]
51. Flugel, E. Mikrofazielle Untersuchungen in der Alpinen Trias: Methoden and Probleme. *Mitt. Ges. Geol. Bergbaustud.* **1972**, *21*, 9–64.
52. Vamvaka, A.; Spiegel, C.; Frisch, W.; Danisik, M. Fission track data from the Mesohellenic Trough and the Pelagonian zone in NW Greece: Cenozoic tectonics and exhumation of source areas. *Int. Geol. Rev.* **2010**, *52*, 223–248. [CrossRef]
53. Rassios, A.; Moores, E. Heterogeneous mantle complex, crustal processes, and obduction kinematics in a unified Pindos-Vourinos ophiolitic slab (northern Greece). *Geol. Soc. Lond. Spec. Publ.* **2006**, *260*, 237–266. [CrossRef]
54. Kilias, A.; Vamvaka, A.; Falalakis, G.; Sfeikos, A.; Papadimitriou, E.; Gkarlaouuni, C.; Karakostas, B. The Mesohellenic Trough and the Paleogene Thrace Basin on the Rhodope Massif, their Structural Evolution and Geotectonic Significance in the Hellenides. *J. Geol. Geosci.* **2015**, *4*, 2. [CrossRef]
55. Brunn, J.H. Contribution à l'étude géologique du Pinde septentrional et d'une partie de la Macédoine occidentale. *Ann. Géol. Pays Hell.* **1956**, *8*, 346–358.
56. Papanikolaou, D.; Lekkas, E.; Mariolakos, H.; Mirkou, R. Contribution on the geodynamic evolution of the Mesohellenic trough. *Bull. Geol. Soc. Greece* **1988**, *20*, 17–36.
57. Aubouin, J. Contribution à l' étude géologique de la Grèce septentrionale: Le confins de l' Epire et de la Thessalie. *Ann. Géol. Pays Hell.* **1959**, *10*, 525.
58. Koukouzas, C.; Koukouzas, N. Coals of Greece: Distribution, quality and reserves. *Geol. Soc. Spec. Publ.* **1995**, *82*, 171–180. [CrossRef]
59. Wilson, M.; Monea, M. IEA GHG Weyburn CO_2 monitoring and storage project: Summary report 2000–2001. In Proceedings of the 7th International Conference Greenhouse Gas Control Technology (GHGT-7), Vancouver, BC, Canada, 5–9 September 2005.
60. Koukouzas, N.; Ziogou, F.; Gemeni, V. Preliminary assessment of CO_2 geological storage opportunities in Greece. *Int. J. Greenh. Gas Control* **2009**, *3*, 502–513. [CrossRef]
61. Tasianas, A.; Koukouzas, N. Assessing the Potential of the Mesohellenic Trough and Other Sites, in Greece, for CO_2 Storage. *Procedia Earth Planet. Sci.* **2015**, *15*, 607–612. [CrossRef]
62. Saccani, E.; Photiades, A. Mid-ocean ridge and supra-subduction affinities in the Pindos ophiolites (Greece): Implications for magma genesis in a forearc setting. *Lithos* **2004**, *73*, 229–253. [CrossRef]
63. Pomonis, P.; Tsikouras, B.; Hatzipanagiotou, K. Geological evolution of the Koziakas ophiolitic complex (W. Thessaly, Greece). *Ofioliti* **2005**, *30*, 77–86.
64. Koutsovitis, P. Gabbroic rocks in ophiolitic occurrences from East Othris, Greece: Petrogenetic processes and geotectonic environment implications. *Mineral. Petrol.* **2011**, *104*, 249–265. [CrossRef]
65. Saccani, E.; Beccaluva, L.; Photiades, A.; Zeda, O. Petrogenesis and tectono-magmatic significance of basalts and mantle peridotites from the Albanian–Greek ophiolites and sub-ophiolitic mélanges. New constraints for the Triassic–Jurassic evolution of the Neo-Tethys in the Dinaride sector. *Lithos* **2011**, *124*, 227–242. [CrossRef]
66. Stouraiti, C.; Pantziris, I.; Vasilatos, C.; Kanellopoulos, C.; Mitropoulos, P.; Pomonis, P.; Moritz, R.; Chiaradia, M. Ophiolitic Remnants from the Upper and Intermediate Structural Unit of the Attic-Cycladic Crystalline Belt (Aegean, Greece): Fingerprinting Geochemical Affinities of Magmatic Precursors. *Geosciences* **2017**, *7*, 14. [CrossRef]
67. Mortazavi, M.; Sparks, R. Origin of rhyolite and rhyodacite lavas and associated mafic inclusions of Cape Akrotiri, Santorini: The role of wet basalt in generating calcalkaline silicic magmas. *Contrib. Mineral. Petrol.* **2004**, *146*, 397–413. [CrossRef]
68. Bachmann, O.; Deering, C.; Ruprecht, J.; Huber, C.; Skopelitis, A.; Schnyder, C. Evolution of silicic magmas in the Kos-Nisyros volcanic center, Greece: A petrological cycle associated with caldera collapse. *Contrib. Mineral. Petrol.* **2012**, *163*, 151–166. [CrossRef]

69. Pe-Piper, G.; Panagos, A.G. Geochemical characteristics of the Triassic volcanic rocks of Evia: Petrogenetic and tectonic implications. *Ofioliti* **1989**, *14*, 33–50.
70. Pe-Piper, G.; Piper, D. Neogene backarc volcanism of the Aegean: New insights into the relationship between magmatism and tectonics. *Geol. Soc. Am.* **2007**, *418*, 17–31. [CrossRef]
71. Fytikas, M.; Innocenti, F.; Manetti, P.; Peccerillo, A.; Mazzuoli, R.; Villari, L. Tertiary to Quaternary evolution of volcanism in the Aegean region: The Geological Evolution of the Eastern Mediterranean. In *Geological Society Special Publication*; Dixon, J.E., Robertson, A.H.F., Eds.; The Geological Society Publishing House: Bath, UK, 1984; Volume 17, pp. 687–699.
72. Innocenti, F.; Agostini, S.; Doglioni, C.; Piero, M.; Tonarini, S. Geodynamic evolution of the Aegean: Constraints from the Plio-Pleistocene volcanism of the Volos-Evia area. *J. Geol. Soc. Lond.* **2010**, *167*, 475–489. [CrossRef]
73. Nicholls, I.A. Petrology of Santorini Volcano, Cyclades, Greece. *J. Petrol.* **1971**, *12*, 67–119. [CrossRef]
74. Nicholls, I.A. Santorini volcano, greece—Tectonic and petrochemical relationships with volcanics of the Aegean region. *Tectonophysics* **1971**, *11*, 377–385. [CrossRef]
75. Bridger, D.W.; Allen, D.M. Designing aquifer thermal energy storage systems. *ASHRAE J.* **2005**, *47*, S32.
76. Vance, D. Heat in Groundwater Systems Part I—The Fundamentals. *Natl. Environ. J.* **1996**, *6*, 30–31.
77. Brandt, M.J.; Johnson, K.M.; Elphinston, A.J.; Ratnayaka, D.D. *Twort's Water Supply*, 7th ed.; Elsevier: Amsterdam, The Netherlands, 2017.
78. Maidment, D.R. *Handbook of Hydrology*; McGraw-Hill: New York, NY, USA, 1992.
79. Price, M. *Introducing Groundwater*, 2nd ed.; Chapman and Hall: London, UK, 1996.
80. Crain, E.R. *Crain's Petrophysical Handbook*; Spectrum 2000 Mindware: Rocky Mountain House, AB, Canada, 1987.
81. Nathenson, M. *Physical Factors Determining the Fraction of Stored Energy Recoverable from Hydrothermal Convection Systems and Conduction-Dominated Areas*; U.S. Geological Survey Open-File Report: Menlo Park, CA, USA, 1975; p. 50.
82. White, D.E.; Williams, D.L. *Assessment of Geothermal Resources of the United Sates*; U.S. Geological Survey Circular: Arlington, VA, USA, 1975; p. 155.
83. Muffler, L.J.P. *Assessment of Geothermal Resources of the United States*; U.S. Geological Survey Circular: Roosevelt, UT, USA, 1979; p. 163.
84. Franco, A.; Donatini, F. Methods for the estimation of the energy stored in geothermal reservoirs. *J. Phys. Conf. Ser.* **2017**, *796*, 012025. [CrossRef]
85. Beardsmore, G.; Rybach, L.; Blackwell, D.; Baron, C. A Protocol for Estimating and Mapping Global EGS Potential. *GRC Trans.* **2010**, *34*, 301–312.
86. Gao, L.; Zhao, J.; An, Q.; Liu, X.; Du, Y. Thermal performance of medium-to-high-temperature aquifer thermal energy storage systems. *Appl. Therm. Eng.* **2019**, *146*, 898–909. [CrossRef]
87. Van Lopik, J.H.; Hartog, N.; Jan Zaadnoordijk, W. The use of salinity contrast for density difference compensation to improve the thermal recovery efficiency in high-temperature aquifer thermal energy storage systems. *Hydrogeol. J.* **2016**, *24*, 1255–1271. [CrossRef]
88. Berger, B.; Anderson, K. *Modern Petroleum: A Basic Primer of the Industry*, 3rd ed.; Pennwell Books: Houston, TX, USA, 1992; p. 517.
89. North, F.K. *Petroleum Geology*; Allen & Unwin: Boston, MA, USA, 1985.
90. Franz, P.; Neville-Lamb, M.; Azwar, L.; Quinao, J. Calculation of Geothermal Stored Heat from a Numerical Model for Reserve Estimation. In Proceedings of the World Geothermal Congress, Melbourne, Australia, 19–25 April 2015.
91. Muffler, P.; Cataldi, R. Methods for regional assessment of geothermal resources. *Geothermics* **1978**, *7*, 53–89. [CrossRef]
92. Schellschmidt, R.; Hurter, S. Atlas of Geothermal Resources in Europe. *Geothermics* **2003**, *32*, 779–787. [CrossRef]
93. Pellegrini, M.; Bloemendal, M.; Hoekstra, N.; Spaak, G.; Andreu Gallego, A.; Rodriguez Comins, J.; Grotenhuis, T.; Picone, S.; Murrell, A.J.; Steeman, H.J. Low carbon heating and cooling by combining various technologies with Aquifer Thermal Energy Storage. *Sci. Total Environ.* **2019**, *665*, 1–10. [CrossRef]

94. Todorov, O.; Alanne, K.; Virtanen, M.; Kosonen, R. A method and analysis of aquifer thermal energy storage (ATES) system for district heating and cooling: A case study in Finland. *Sustain. Cities Soc.* **2020**, *53*, 101977. [CrossRef]
95. Hooimeijer, F.; Maring, L. The significance of the subsurface in urban renewal. *IRPUS* **2018**, *11*, 303–328. [CrossRef]
96. Kostakis, I. Socio-demographic determinants of household electricity consumption: Evidence from Greece using quantile regression analysis. *CRSUST* **2020**, in press. [CrossRef]
97. Azeiteiro, U.; Davim, J. *Higher Education and Sustainability*, 1st ed.; CRC Press: Boca Raton, FL, USA, 2020.
98. Tichler, R.; Bauer, S. Power to gas. In *Storing Energy with Special Reference to Renewable Energy Sources*; Letcher, T.M., Ed.; Elsevier: Oxford, UK, 2016; p. 590.
99. Renpu, W. Basis of Well Completion Engineering. In *Advanced Well Completion Engineering*, 3rd ed.; Renpu, W., Ed.; Gulf Professional Publishing: Oxford, UK, 2011; pp. 1–74.
100. Speight, J.G. Recovery, storage, and transportation. In *Natural Gas*, 2nd ed.; Speight, J.G., Ed.; Gulf Professional Publishing: Boston, MA, USA, 2019; pp. 149–186.
101. Speight, J.G. (Ed.) Reservoirs and Reservoir Fluids. In *Handbook of Hydraulic Fracturing*; Wiley: Hoboken, NJ, USA, 2016; pp. 27–54.
102. Papadopoulou, D.; Tourkolias, C.N.; Mirasgedis, S. Assessing the macroeconomic effect of gas pipeline projects: The case of Trans-Adriatic Pipeline on Greece. *SPOUDAI* **2015**, *65*, 100–118.
103. Khalova, G.O.; Illeritskiy, N.I.; Smirnova, V.A. Prospects for the Construction of the Poseidon Gas Pipeline as a Factor in Supplying the Needs of the Southern Europe Countries with Natural Gas. *Int. J. Energy Econom. Policy* **2019**, *9*, 143–148.
104. Kotek, P.; Granado, P.C.d.; Egging, R.; Toth, B.T. European Natural Gas Infrastructure in the Energy Transition. In Proceedings of the 16th International Conference on the European Energy Market (EEM), Ljubljana, Slovenia, 18–20 September 2019; pp. 1–6.
105. Kitsilis, M.-C. Issues for Underground Gas Storage (UGS) in 'South Kavala' offshore gas field. In Proceedings of the 5th South East Europe Energy Dialogue, Thessaloniki, Greece, 2–3 June 2011.
106. Ozarslan, A. Large-scale hydrogen energy storage in salt caverns. *Int. J. Hydrog. Energy* **2012**, *37*, 14265–14277. [CrossRef]
107. Winter, C.-J.; Nitch, J. *Hydrogen as an Energy Carrier: Technologies, Systems, Economy*; Spiringer: Stuttgart, Germany, 1988.
108. Eiken, O.; Ringrose, P.; Hermanrud, C.; Nazarian, B.; Torp, T.A.; Høier, L. Lessons learned from 14 years of CCS operations: Sleipner, In Salah and Snøhvit. *Energy Procedia* **2011**, *4*, 5541–5548. [CrossRef]
109. Beaubien, S.E.; Jones, D.G.; Gal, F.; Barkwith, A.K.A.P.; Braibant, G.; Baubron, J.-C.; Ciotoli, G.; Graziani, S.; Lister, T.R.; Lombardi, S.; et al. Monitoring of near-surface gas geochemistry at the Weyburn, Canada, CO_2-EOR site, 2001–2011. *Int. J. Greenh. Gas Control* **2013**, *16*, S236–S262. [CrossRef]
110. Andritsos, N.; Arvanitis, A.; Papachristou, M.; Fytikas, M.; Dalabakis, P. Geothermal Activities in Greece During 2005-2009. In Proceedings of the World Geothermal Congress Bali, Bali, Indonesia, 25–29 April 2010; pp. 25–29.
111. Jin, C.; Liu, L.; Yiman, l.; Zeng, R. Capacity assessment of CO_2 storage in deep saline aquifers by mineral trapping and the implications for Songliao Basin, Northeast China. *Energy Sci. Eng.* **2017**, *5*, 81–89. [CrossRef]
112. Hannis, S.; Lu, J.; Chadwick, A.; Hovorka, S.; Kirk, K.; Romanak, K.; Pearce, J. CO_2 storage in depleted or depleting oil and gas fields: What can we learn from existing projects? *Energy Procedia* **2017**, *114*, 5680–5690. [CrossRef]

Article

Petrographic Characteristics of Sandstones as a Basis to Evaluate Their Suitability in Construction and Energy Storage Applications. A Case Study from Klepa Nafpaktias (Central Western Greece)

Petros Petrounias [1,*], Panagiota P. Giannakopoulou [1], Aikaterini Rogkala [1], Maria Kalpogiannaki [1], Petros Koutsovitis [1], Maria-Elli Damoulianou [1] and Nikolaos Koukouzas [2]

1 Section of Earth Materials, Department of Geology, University of Patras, 26504 Patras, Greece; peny_giannakopoulou@windowslive.com (P.P.G.); krogkala@upatras.gr (A.R.); maria.kalpogiannaki@gmail.com (M.K.); pkoutsovitis@upatras.gr (P.K.); marilliedam@gmail.com (M.-E.D.)

2 Chemical Process & Energy Resources Institute, Centre for Research & Technology Hellas (CERTH), Maroussi, 15125 Athens, Greece; koukouzas@certh.gr

* Correspondence: Geo.plan@outlook.com

Received: 4 February 2020; Accepted: 27 February 2020; Published: 2 March 2020

Abstract: This study investigates how the petrographic features of Klepa Nafpaktias sandstones affect their behavior in construction applications such as concrete, in environmental applications such as energy storage as well as whether they are suitable for the above uses. For achieving this goal, sandstones (ten samples) were collected in order to study their petrographic characteristics using petrographic microscope and GIS software, as well as their basic physical, mechanical and physicochemical properties were also examined. Concrete specimens (C25/30) were made according to international standards including the investigated aggregate rocks in various grain sizes. Various sandstones were tested and classified in three district groups according to their physicomechanical features as well as to their petrographic and microtopographic characteristics. Concrete strength's results determined the samples into three groups which are in accordance with their initial classification which was relative to their grain size (coarse to fine-grained). As the grain size decreases their physicomechanical and physicochemical properties get better resulting in higher concrete strength values (25 to 32 MPa). Furthermore, the proposed ratio *C/A* (crystals/mm^2) seems to influence the aggregate properties which constitute critical factors for the final concrete strength, presenting the more fine-grained sandstones as the most suitable for concrete aggregates. Concerning the use of Klepa Nafpaktias sandstones as potential energy reservoirs, the studied sandstones presented as suitable for CO_2 storage according to their physicomechanical characteristics.

Keywords: petrographic characteristics; sandstones; physicomechanical properties; concrete petrography; CO_2 storage

---

1. Introduction

Applied petrography constitutes an essential tool for the assessment of natural rocks or recycling materials for different useful applications such as concrete and energy storage. Petrography, generally, using a combination of methods such as microscopic observations (polarizing and scanning electron microscope) and chemical analysis examines the nature of each given rock/material showing the main relationship of texture, structure, composition and alteration degree [1–9]. Through these relationships, petrography may explain the physicomechanical and physicochemical properties of materials/rocks as well as the relationships among them. It is well-known that the already above-mentioned properties are the critical ones that define the particular use of each given material/rock either construction or environmental applications.

Concrete is one of the most important and useful composite material, which is made from a mixture of cement, aggregates, water and sometimes admixtures in required proportions [10–13]. Cement and water bonded with aggregates constitute the concrete. Aggregates constitute the main components of concrete and occupy between 70% and 80% of the concrete volume [10]. Nevertheless, concrete performance depends on the aggregate particles quality [11–14]. Crushed rocks derived from various geological sources are the natural aggregates [15]. Critical factors for the suitability of aggregates in construction use are their physicomechanical behavior. This behavior is affected by the mineralogical and textural features of aggregates which play critical role in its strength and therefore in concrete strength. The most common types of rocks used in concrete production are classified into igneous, sedimentary and metamorphic rocks. Aggregates can be expected to have an important influence on the concrete's properties [16]. Such rocks are mainly limestone, granite, sandstone, quartzite, dolomite, marble, dacite etc. Each of these rock types is more or less suitable for uses as concrete aggregates, based on their petrographic characteristics and therefore on physicomechanical properties which contribute to reinforcing the strength of the concrete.

Sandstone is a widespread aggregate presenting various construction applications such as concrete. The physicomechanical properties of the aforementioned rock lithotype are quite different and aggregates such as quartzite, subarcose and greywacke can produce various behaviors of hardened concrete. Thus, it can be seen that these sandstone aggregates can obviously be characterized to acquire relatively predictable aggregate and concrete performances [17]. Sandstone used as aggregate of variable sizes within concrete may result in influencing the corresponding strength. In addition, it is regarded as significant for these aggregates to be graded when applied for concrete production. Moreover, it is well-known that sandstone is vastly affected by moisture content which results in the decrease of the identified mechanical property behavior in brittle construction material. Additionally, these sedimentary rocks have the tendency to display smaller strength compared to conventional aggregate material such as limestones. Sandstone performance behavior under anhydrous conditions is regarded as being good, which is not in the case of hydrous conditions since it is regarded as poor behavior especially in sandstones that are not well cemented [18,19]. Quartz content in concrete prepared by sandstone aggregates determines the concrete application [20]. Yilmaz and Tugrul [21] reported that concretes produced by comparable qualities and quantities cements exhibit variability in strength values depending on: mineralogy assemblage of aggregate, type of cement, textural and physico-mechanic characteristics.

Many researchers have investigated the correlations between the percentages of specific mineralogical compositions of aggregates and the final compressive concrete strength. Petrounias et al. [22,23] when investigating igneous rocks from Greece concluded that the alteration products of serpentinite-bearing rocks and andesitic-intermediate rocks have a profound impact on their mechanical behavior, that apparently affect their ability to be characterized as suitable concrete aggregate material. On the other hand, Yılmaz and Tugrul [21] evaluated sandstone aggregates from Turkey based on the parameters that include: physical properties, mechanical properties and compositional variations. They concluded that concretes produced by comparable qualities and quantities of cements present variabilities in the obtained strength values regarding mineralogy assemblage, the textural and physico-mechanic characteristics of the tested sandstones used in the size of aggregate.

During the last decades, only a few attempts to combine the database and visualization facilities of Geographic Information System (GIS) software and petrographic features of rocks have been carried out. In these studies, polarizing microscope images have been used in order to identify and visualize rock textures on microscopic scale. Li et al. [24] using GIS software to segment and analyze boundaries in the perimeters of grains, proposed a procedure to be applied on examined samples. Barraud [25] has used GIS techniques to process vectorized textural features, whereas Fernandez et al. [26] used the same application to parametrize single grain crystals through map development methods. An innovative methodology combining GIS and petrographic characteristics of various rocks has been applied by Tarquini [27].

Economic growth and a rising global population means that the worldwide demand of energy will be rising with very fast pace. This increases concerns that the extensive use of fossil-fuels should be mitigated, allowing space for further development of renewable energy sources. The problem that arises with the use of the latter is that most of these sources are intermittent and therefore energy storage applications are necessary to make them available around-the-clock for uninterrupted power supply [28]. Suitable subsurface geological formations can serve as energy storage reservoirs depending on the storage purpose and the type of energy source. Energy storage systems include that of thermal energy, CO_2, compressed air, hydrogen storage, natural gas and underground pumping of water.

Rocks consisting geological formations must fulfill certain criteria to be considered as a candidate reservoir for potential thermal-energy storage (TES), compressed air energy storage (CAES) and carbon capture and storage (CCS) applications. These criteria have been noted by researchers (e.g., [28–30]) stating that rocks should display high values of thermal conductivity, specific heat capacity, and density to enable high storage efficiency. Low porosity values correlate positively with high values of bulk density and uniaxial compressive strength, which are necessary to ensure not only the optimum energy storage criteria but to avoid fracture development and disintegration [28]. Research conducted by Allen et al. [31] and Tiskatine et al. [32] suggest that formations consisting of sandstones may serve as proper energy storage reservoirs, provided that they meet compositional (e.g., calcium-or silica-rich), textural and structural and also not having been significantly affected secondary alteration processes.

Aim of this study is to highlight the effect of petrographic characteristics of sandstones from Klepa Nafpaktias (central western Greece) as a decisive factor in the final strength of the produced concrete specimens by sandstones aggregates and also to examine their potential use as geological reservoir for carbon capture and storage (CCS) applications.

2. Geological Setting

The study area is Klepa Nafpaktias that geographically belongs to the regional unit of Aitolia and Akarnania and geologically to the Pindos Geotectonic Zone, which comprises an intricate thrust belt with allochthonous Mesozoic and Tertiary deep-water tectono-stratigraphic units [33], which are developing along the central Western Greece (Figure 1) and extend into Albania and former Yugoslavia to the north [34] and Crete, Rhodes [35] and Turkey [36,37], toward the south and southeast. The Pindos Zone sedimentary sequence was deposited in an extended oceanic rift related basin that likely was created during the Middle Triassic era, located in the Apulia extensive marginal platform [38] and more specifically in the Gavrovo-Tripolis sedimentary platform [33,39] that emerged periodically, and now lies westwards of Pindos, and the Pelagonian continental block in the east [40–43]. The progressive closure of the Pindos oceanic basin were initiated during the end of Maastrichtian, as recorded by the gradual alteration, from predominantly carbonates intercalating with radiolaria to siliciclastic/turbiditic lithofacies (Paleocene flysch deposition) derived from the north and east sectors [33,39,44]. The complete closure of the Pindos Ocean during the Eocene led to the detachment of the deep-sea sedimentary cover from the accretionary oceanic prism, which was later overthrusted in a westward direction onto the Pelagonian geotectonic platform, forming multiple layers of thrust sheets [38,42].

The sedimentary alternating strata of Pindos Zone consist of deep-water carbonate, siliciclastic and siliceous rocks of Late Triassic to Eocene age [38,39,42], mainly including the following units (Figure 1): (1) Ophiolite complex of Pindos (Jurassic age); (2) Limestones of Orliakas (Late Cretaceous age); (3) A Mélange formation, namely that of the Avdella (Late Triassic to Late Jurassic age), (4) the Sediments of deep water type, namely that of Dio Dendra (Late Jurassic to Late Cretaceous age); and (5) a thick flysch formation assigned to the Pindos zone (Late Cretaceous to Tertiary age) according to Jones et al. [45]. More extensively, the Pindos flysch consists of thin- to thick-bedded sandstones and mudstones in alteration with marly-oolitic limestones and cherts (reference). According to Konstantopoulos and

Zelilidis [46], the sandstones of the Pindos Flysch were possibly cited within an accretionary prism located close to an active marginal basin, supplied by predominantly basic/ultra-basic and less felsic material. Furthermore, Faupl et al. [47] conducted a heavy mineral examination, suggesting that the clastic material of the Pindos flysch has an eastwards origin, while a petrographic and geochemical study by Vakalas et al. [48] on sandstone samples from Epirus and Akarnania regions suggests a granitic source and a supply from the Pelagonian Zone correspondingly. Detrital modes of sandstone suites reveal the lithological composition of source terranes and the tectonostratigraphic level reached by erosion in space and time.

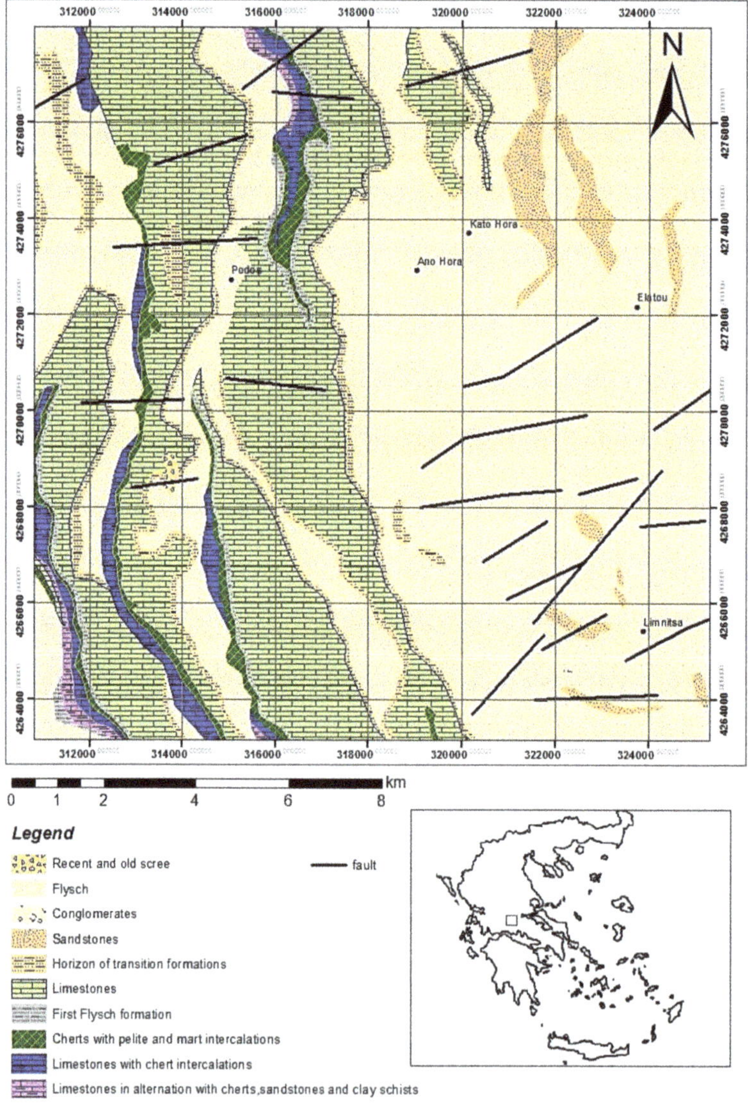

Figure 1. Geological map of the Klepa Nafpaktias [49] (Central Western Greece) region (modified after fieldwork mapping by using ArcMap 10.1).

3. Materials and Methods

3.1. Materials

Ten samples from different type of sandstones (coarse-grained and fine-grained) were collected from the studied area. These samples were tested for their petrographic characteristics, physicomechanical and physicochemical properties in order to be classified for their suitability as concrete aggregates. Normal Portland cement (CEM II 32.5N) was used in this study, which conformed to EN 197-1 [50] standard with the collected aggregates in order to produce concrete. For the mixing as well as for the curing of concrete, a potable water with pH = 7, characterized as free of clay, salt, organic matter and silt was chosen to be used. The same volume of aggregate per m^3 of the mixture was retained for keeping consistent composition in all the produced specimens. The proportions used for cement, aggregates and water ratio were those of 1/6/0.63. The same collected sandstones were also investigated for their potential use as reservoir for thermal-energy storage (TES) and compressed air energy storage (CAES) applications.

3.2. Methods

3.2.1. Rock Material Tests

Polished thin sections of the collected sandstones were first tested via a petrographic microscope (EN 932-3 [51]) for identifying their mineralogical and textural characteristics. The used microscope was of the type of Leitz Ortholux II POL-BK Ltd., Midland, ON, Canada one. In a further stage, their petrographic characteristics were investigated as well as the quantification of their mineralogical composition was calculated using the ArcMap 10.1 software, in which six representative thin sections of the studied groups (two per sample) were investigated.

Secondary electron images (SEI) were used for the identification of the microtopography of the aggregates (BS 812 Part 1 [52], whereas six qualitative categories were outlined (porous, honeycomb, crystalline, rough, granular, smooth and glassy).

A hammer was used in order to crush the studied sandstones into smaller pieces. Then, a laboratory jaw crusher was used for extra fractioning of the samples. For the preparation of the cylindrical specimens with diameters varying from 50 to 54 mm and with ratio of length to diameter varying from 2.2 to 2.4 mm, laboratory core drill combined with saw machines were used.

The investigated physicomechanical and physicochemical properties of this study were the total porosity (n_t) (ISRM 1981 standard [53]), the magnesium sulfate ($MgSO_4$) (S) (EN 1367-2 [54]), the water absorption (w_a) (EN 1097-6 [55]), the resistance in abrasion and attrition-Los Angeles (LA) (ASTM C-131 [56]), and the uniaxial compressive strength (UCS) (ASTM D 2938-95 [57]), properties crucial for further quality of the aggregates.

3.2.2. Concrete Tests

Twenty normal concrete cube specimens having dimensions of 150 × 150 mm were produced by the ten investigated sandstone aggregates (ACI-211.1-91) (Table 1) [58]. For all the produced concrete specimens, the used parameters retained fixed. The aggregates were crushed and sieved via standard sieves and separated to achieve the sizes: 2.00–4.75, 4.45–9.5 and 9.5–19.1 mm. After 24 h, the samples were transferred from the mold and for 28 days they were cured in water at 20 ± 3 °C. A compressive testing machine, having an increasing rate of load of 140 kg/cm^2 per minute, was used in order to calculate the concrete strength by the division of the value of the load at the moment of failure over the area of specimen. The concrete strength was calculated via BS EN 12390-3:2009 standard [59].

In this stage, the examination of the concrete textural features was carried out when using polished thin sections in a petrographic microscope (ASTM C856–17) [60]. A 3D depiction of the petrographic characteristics of the concrete as well as of the studied sandstone aggregates was carried out by the 3D Builder software using thin sections.

Table 1. Quantification of the modal composition of the representative investigated groups of sandstones.

		Modal Composition							Ratio
	Samples	**Quartz**	**K-Feldspars**	**Plagioclase**	**Calcite**	**Muscovite**	**Total Cement**		***C/A***
Group I	K.L5	24.96	29.20	0.53	1.43	1.60	42.28		11.61
	K.L9	26.00	28.34	0.51	1.43	1.58	42.14		11.59
Group II	K.L1	25.56	16.82	0.50	8.40	1.37	47.35		21.40
	K.L3	25.52	16.81	0.65	8.05	1.33	47.64		20.80
Group III	K.L7	29.50	6.64	0.20	2.46	4.12	57.08		56.50
	K.L10	29.10	6.62	0.26	2.44	1.41	57.44		55.70

4. Results

4.1. Test Results of Aggregates

4.1.1. Petrographic Features of Aggregates Using Petrographic Microscope

The studied sandstones derived from Klepa area have been divided according to the petrographic analysis into three district groups. These groups are based on the grain size of the collected sandstones and they were characterized as coarse to fine-grained ones.

Group I: Coarse-grained sandstones (KL.5, KL.9)

These sandstones comprise sub-angular to angular grains (Figure 2a,b). They are generally moderate to poor sorted. The mineralogical composition mainly includes quartz, K-feldspars, plagioclase, calcite, mica and in minor amounts muscovite, chlorite and biotite as well as lithic fragments (Table 1). These sandstones present mainly siliceous cement. Quartz is mostly displayed as undulose monocrystalline and less as polycrystalline grains. The monocrystalline quartz grains vary from sub-angular to angular, while the polycrystallines range from sub-angular to sub-round. The grain contacts presented as straight to suture. K-feldspars grains vary in size, from small to large with euhedral to subhedral shape, whereas plagioclase is observed in smaller grains. In general, the fragments are sub-rounded and sub-angular to angular and they are mainly comprised of clasts of quartz, feldspars as well as by rock-fragments of basalt and gabbro. Traces of carbonate fossils are also observed in several samples (e.g., KL.5).

Group II: Medium-grained sandstones (KL.1, KL.2, KL.3, KL.6)

The medium-grained sandstones can be classified as quartz sandstones. They are moderately sorted and their grains are sub-angular to sub-round. The main mineralogical composition includes quartz which forms monocrystalline and polycrystalline grains, K-feldspars, calcite and muscovite (Figure 2c). Polycrystalline quartz shows interlocking texture. Feldspars (mainly microcline) are presented in lesser amounts, including the weathered varieties (Table 1). Cement is mainly siliceous and locally calcareous.

Group III: Fine-grained sandstones (KL.4, KL.7, KL.8, KL.10)

In the fine-grained quartz sandstones, framework grains are mainly sub-angular to sub-round. They are characterized as well sorted quartz sandstones. The modal composition mostly comprises of quartz, K-feldspars, calcite, and mica (mainly muscovite) which is presented in bigger amounts in contrast to the other two groups. Cementing material is mainly siliceous (Table 1).

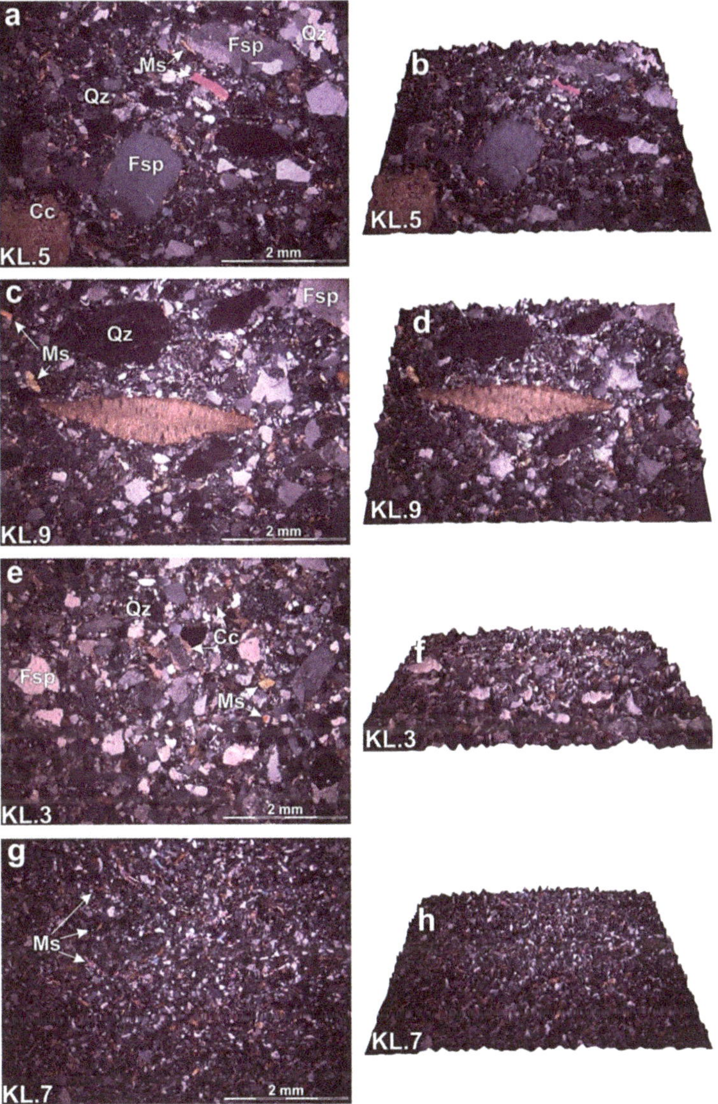

Figure 2. Photomicrograph of textural characteristics of sandstone aggregates (Nicols+) and 3D depiction of the studied sandstones respectively: (**a**) clastic texture presented in a coarse-grained quartz sandstone with quartz (Qz), K-feldspars (K-Fs), plagioclase (Plg), muscovite (Ms) and calcite (Cc); (**b**) 3D depiction of coarse-grained sandstone; (**c**) clastic texture presented in a coarse-grained sandstone containing large particles of carbonate fossils; (**d**) 3D depiction of coarse grained sandstone; (**e**) clastic texture presented in a medium-grained quartz sandstone with quartz (Qz), K-feldspars (K-Fs) and calcite (Cc); (**f**) 3D depiction of coarse-grained sandstone; (**g**) clastic texture presented in a fine-grained quartz sandstone with muscovite (Ms); (**h**) 3D depiction of fine-grained sandstone.

4.1.2. Petrographic Features of Aggregates Using GIS Method

In this paper, GIS method was used as a new approach for petrographic analysis of the investigated sandstones. For this reason, six thin sections, representative of the investigated sandstones (two sections for each group) were used in order to be analyzed via GIS method. More specifically, a part of the same size and in the same site of the thin section has been chosen to be digitized via ArcMap 10.1 software. Each digitized polygon corresponds to a different grain of the sandstone. The result of that process is the creation of a map that shows the modal composition of the tested rocks as well as their textural characteristics such as the grain size (Figure 3). In a next stage, the semi-quantification of the mineralogical composition of the studied sandstones was carried out, showing that Group III presents higher content of quartz than the other two groups, Group I displays intense higher content of feldspars in contrast to Group II and III (Table 1), while Group II is presented as more enriched in calcite (Table 1). Furthermore, Group III displays significant high content of muscovite. Concerning the containing cement, Group III presents higher content of silica cement in contrary to the other two groups (Table 1). After the petrographic analysis via the GIS proposed method, the ratio *C/A* was calculated (Table 1). *C/A* (crystals/mm^2) is the ratio of the sum of the measured crystals to the measured area (mm^2). As can be seen in Table 1, Group I, which contains the coarser grains, presents an average of *C/A* 11.60 in contrast to samples of Group III which presents values of *C/A* from 55.70 to 56.50 and this group consists of the smallest size grains.

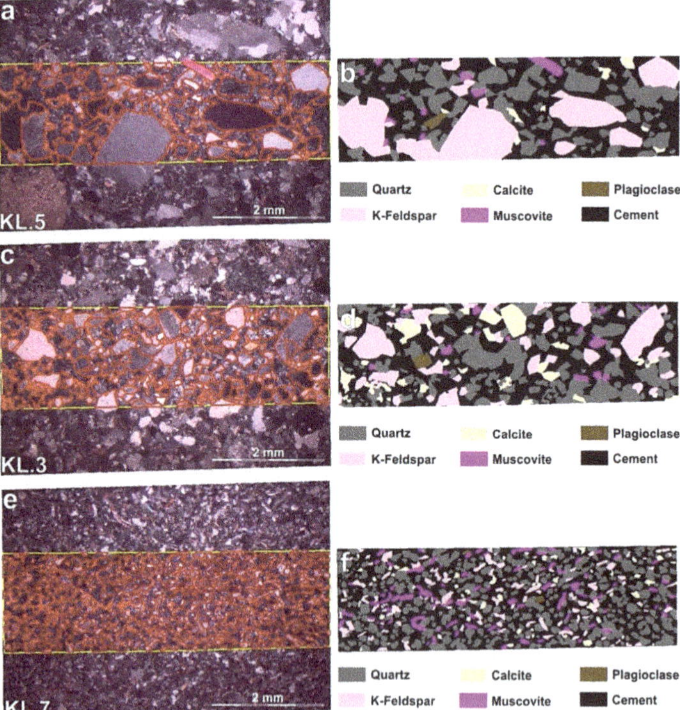

Figure 3. Representative images from the studied groups derived from the ArcMap 10.1 software showing: (**a**) The part of the thin section of the coarse-grained sandstone which has been analyzed; (**b**) the output after the digitalization of the investigated part of each thin section where each mineral phase has been attributed by different colors; (**c**) the part of the thin section of the medium-grained sandstone which has been analyzed; (**d**) the output after the digitalization of the investigated part of each thin section where each mineral phase has been attributed by different colors; (**e**) the part of the thin section of the fine-grained sandstone which has been analyzed; (**f**) the output after the digitalization of the investigated part of each thin section where each mineral phase has been attributed by different colors.

4.1.3. Microtopographic Characteristics of the Tested Sandstones

The surface textural characteristics of the examined sandstone particles were used to classify the quartz sandstones in groups consistent with the above mentioned Groups I to III, as it is shown in Figure 4. The abundance of the coarse size grains of quartz, feldspars and carbonate fossils in the poor sorted sandstone (Group I) is responsible for the smooth surfaces. Sandstones of Group, II constituting medium-grained rocks, are characterized by rough surface texture. The surface of Group III samples can be characterized as rougher in contrast to the other two groups because of the existence of higher content of evenly distributed mica and quartz expressing topographic low areas combined with feldspars which express lower topographic areas.

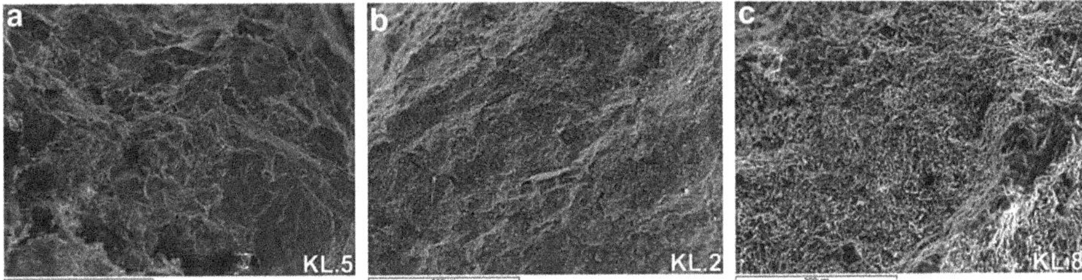

Figure 4. Secondary electron images (SEI) depicting the microtopography of representative sandstone samples observed to their textural and mineralogical characteristics: (**a**) coarse-grained; (**b**) medium-grained; (**c**) fine-grained sandstone.

4.1.4. Physicomechanical Properties of Aggregates

Three distinct groups of sandstones were determined following the results of their physicomechanical properties (Table 2). The calculated mechanical as well as the calculated physical results of the examined rocks present quite wide range. These results have classified the investigated sandstones into groups which are in agreement with those derived from their petrographic study. Group I includes coarse-grained sandstones, which presented the worst values of mechanical properties among all the studied rock samples. Among the examined sandstone aggregates of Group I, sample KL.5, owing to its lower content in quartz, led to a lowest value of total porosity (n_t) as well as in resistance in abrasion. Group II was composed of medium-grained sandstones presented variable physicomechanical performance relative to their mineralogical composition. Group III included fine-grained sandstones, which displayed high physicomechanical parameters among all the determined groups. The fine-grained sandstones, such as KL.4 and KL.8, presented better mechanical characteristics against the coarse-grained sandstones such as KL.5 and KL.9.

Table 2. The results of the physicomechanical properties of the tested sandstones.

Samples	Lithotype	Particle Size	Los Angeles (LA %)	Uniaxial Compressive Strength of Rocks (UCS MPa)	n_t (%)	W_a (%)	S (%)
KL.1	Sandstone	Medium grained (Group II)	20.0	115.0	4.50	2.10	20.00
KL.2	Sandstone	Medium grained (Group II)	21.0	105.0	4.80	1.80	18.00
KL.3	Sandstone	Medium grained (Group II)	22.0	89.0	5.30	2.21	19.00

Table 2. *Cont.*

Samples	Lithotype	Particle Size	Los Angeles (LA %)	Uniaxial Compressive Strength of Rocks (UCS MPa)	n_t (%)	W_a (%)	S (%)
KL.4	Sandstone	Fine grained (Group III)	16.0	112.0	3.50	1.55	13.00
KL.5	Sandstone	Coarse grained (Group I)	29.0	77.0	9.50	3.30	48.00
KL.6	Sandstone	Medium grained (Group II)	19.0	105.0	3.70	2.18	17.00
KL.7	Sandstone	Fine grained (Group III)	13.0	115.0	2.30	0.90	15.00
KL.8	Sandstone	Fine grained (Group III)	15.0	113.0	2.90	1.50	12.00
KL.9	Sandstone	Coarse grained (Group I)	33.0	75.0	19.50	2.80	38.00
KL.10	Sandstone	Fine grained (Group III)	15.0	112.0	3.10	1.60	20.00

4.2. *Concrete Test Results*

4.2.1. Uniaxial Compressive Strength of Concrete

Uniaxial compressive strength test was carried out for the produced specimens, whereas their results can be observed in Table 3. These values varied from 24 to 32 MPa after 28 days of curing. The concrete strength results are in relevant accordance with the strength of their aggregates. Concrete specimens displayed the lowest strength values that have been made by sandstones of Group I as aggregate particles (Tables 2 and 3). More specifically, the coarse-grained sandstone that contains carbonate fossils of big size displays the worst strength value (24 MPa), value which is below the permitted limit for the concrete class C25/30. The concrete specimens, made by medium grained aggregates from Group II, displayed variety on strength values (26 to 31 MPa), whereas on the other hand those made with the finer grained aggregates from Group III showed the highest compressive strength values (30 to 32 MPa).

Table 3. Uniaxial compressive strength test results of the concrete specimens.

Samples	Lithotype	Particle Size	Uniaxial Compressive Strength of Concrete Specimens (UCS_{con} (MPa))
KL.1	Sandstone	Medium grained (Group II)	31.0
KL.2	Sandstone	Medium grained (Group II)	28.0
KL.3	Sandstone	Medium grained (Group II)	26.0
KL.4	Sandstone	Fine grained (Group III)	30.0
KL.5	Sandstone	Coarse grained (Group I)	25.0
KL.6	Sandstone	Medium grained (Group II)	27.0
KL.7	Sandstone	Fine grained (Group III)	32.0
KL.8	Sandstone	Fine grained (Group III)	32.0
KL.9	Sandstone	Coarse grained (Group I)	24.0
KL.10	Sandstone	Fine grained (Group III)	31.0

4.2.2. Petrographic Characteristics of the Investigated Concretes

Careful microscopic analysis of the thin sections of the tested concrete specimens which were studied by using polarizing microscope as well as through the 3D depiction of the same thin sections showed, in general, satisfied cohesion between the aggregate particles and the cement paste among all the concrete specimens produced by the coarse-grained, the medium-grained as well as the fine-grained sandstones (Figure 5). The existence of intense content of silica cement may enhance the bonding between the sandstone aggregates and the cement paste. Even neither in concrete specimens produced by aggregates of Group I nor in those produced by aggregates of Group II and III, loss of material can be observed nor extensive interaction zones.

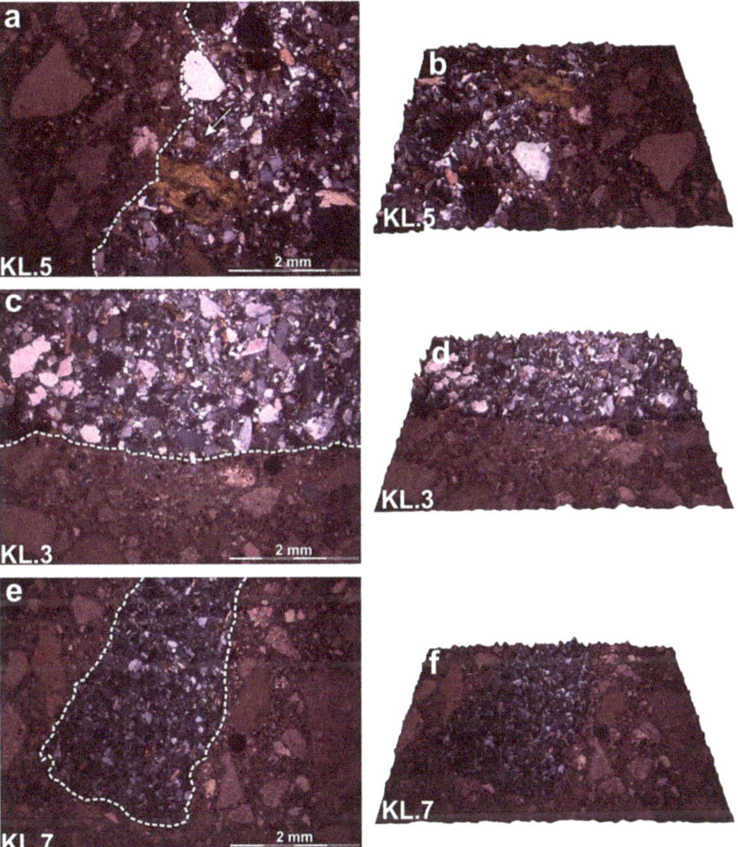

Figure 5. Photomicrographs of representative tested concrete specimens produced by: (**a**) Coarse-grained sandstone aggregate; (**c**) medium-grained sandstone aggregate; (**e**) fine-grained sandstone aggregate and 3D depiction of representative tested concrete specimens produced by: (**b**) coarse-grained sandstone aggregate; (**d**) medium-grained sandstone aggregate; (**f**) fine- grained sandstone aggregate.

5. Discussion

5.1. *The Impact of Petrographic Characteristics on the Sandstone Aggregate Properties and on the Quality of Concrete*

Petrographic characteristics such as mineralogical composition, texture, particle size, alteration and weathering degree of rocks which are used as aggregate materials, constitutes the main factors influencing their properties that are critical for their suitability in various construction and industrial applications [61]. Numerous researchers have correlated physical and mechanical properties of rocks used as aggregates [62–64] giving clear interpretations of the relationships between them which are based on the type of the contained minerals and on their size. Petrounias et al. [65] have proved that the type of the secondary phyllosilicate minerals contained in mafic, ultramafic and intermediate and acidic volcanic rocks is the critical factor that predominately determines the physico mechanical properties of the studied rocks. Additionally, Tugrul and Zarif [66] have reported that the mean grain size is presented as a primary factor influencing the mechanical behavior of the granites which are used as concrete aggregates. More specifically, they have proved that as the mean grain size decreases, the strength of the rock increases respectively. The statistical method that is widely used to determine the relationships between different engineering parameters of rocks is the regression analysis [66–68]. In this paper, where sandstones from Klepa Nafpaktias were studied, strong relationships between the physical and mechanical parameters as well as among mechanical, physical, and physicochemical ones were observed using regression analysis. These correlations as they presented in Figure 6 are mainly dependent on the grain size and lesser on the mineralogical composition and on the amount of the cement. The diagrams of Figure 6 indicates that as the grain size of the investigated sandstone increases, the values of their physical properties increase while the values of their mechanical properties decrease respectively. For example from the diagram of Figure 6a we can observe that Group I, as it is classified after petrographic analysis through the petrographic microscope and verified after the new proposed petrographic analysis via GIS method and is characterized as the more coarse-grained group, presents ratio *C/A* 11.60 on average and higher values of porosity (n_t) (Table 2) and lower resistance in abrasion and attrition (LA). Likewise in the diagram of Figure 6b Group I presented as more capable to absorb water (w_a) and with lower values of uniaxial compressive strength (UCS). Diagrams of Figure 6a,b show the interaction between the physical and mechanical properties which are directly depended on the grain size of the similar mineralogical composition tested sandstones. The lower value of the mechanical strength of the coarse-grained sandstones may be a result of the low, and maybe because of the microtopography, internal attrition between the grains combined with the small percentage of cement, which leads to small angles of attrition relative to the density.

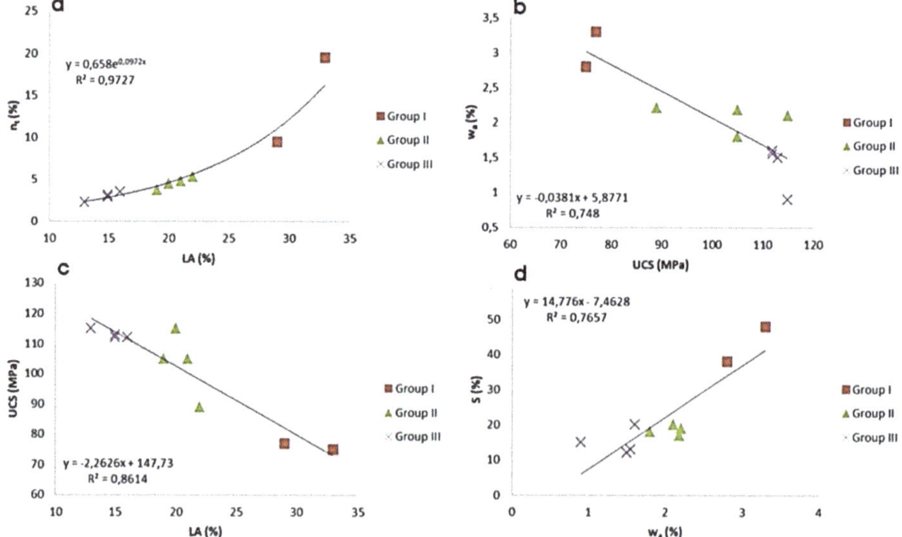

Figure 6. (**a**) LA vs. total porosity (n_t) of rocks; (**b**) compressive strength (UCS) vs. water absorption (w_a) of rocks; (**c**) LA vs. the compressive strength (UCS) of rocks; (**d**) water absorption (w_a) vs. the soundness test (S) of rocks.

In contrast to the comparatively finer grained sandstones, molecular internal forces are developed during the loading presenting better cohesion and bonding among the grains. Furthermore, the porosity (n_t) as well as the water absorption (w_a) seems to significantly be increased in the coarse sandstone rocks against the fine ones, which indicates that larger grains exhibiting weaker cohesion in contrast to the smaller are capable to adsorb more percentage of water around each grain mainly in the form of a surface layer. The diagram in Figure 6c illustrates the interaction of the mechanical properties LA and UCS directly dependent on the grain size. Several researchers have also reported similar relationships between these properties [23,65,69,70] when studying various types of rocks. In this diagram, it seems obvious that the mechanical characteristics of sandstones vary in a similar way under different type of mechanical loadings. For example, rocks of Group I (coarse-grained) present lower resistance in abrasion and attrition and lower compressive strength in contrast to those of Group III (fine-grained) which presented more resistance in abrasion and attrition and with higher strength values. In the diagram of Figure 6d, the relationship between the Soundness test (S) and the water absorption (w_a) is presented, the trend of which is similar to other reported relationships between the Soundness test and physical properties by several researchers [61]. The interpretation given above regarding the ability of coarse-grained sandstones to adsorb water in their structure in contrast to the fine-grained sandstones has a strong effect on their resistance to temperature changes. All of the above theories regarding the effect of grain size on the physicomechanical properties of rocks are quantified and presented below in Figure 7. More specifically, the quantification of the number of minerals per mm^2 (*C/A*) (calculated via GIS) seems to be strongly correlated with the physicomechanical properties of the sandstones. In Figure 7a, it can be seen that as the *C/A* increases, the strength of the rocks increases (Figure 7a) and their resistance in abrasion and attrition increases respectively (Figure 7b), whereas the number of minerals per mm^2 increases as their porosity decreases (Figure 7c). It is noticeable that the above mentioned relationships display high coefficient of correlation (R^2 = 0.72 and R^2 = 0.71) (Table 4) a fact that shows that the grain size constitutes the principal but not the unique factor that influences these properties. This happens because the mineralogical composition of rocks also determines their physicomechanical properties.

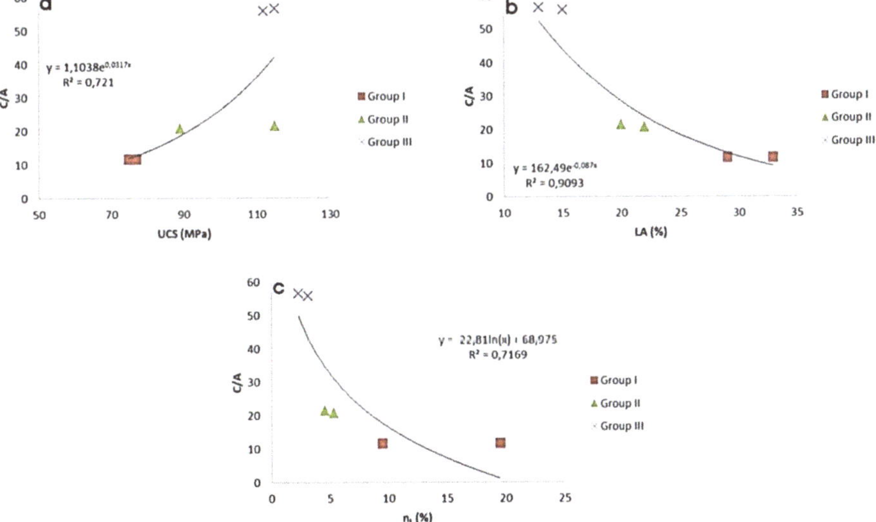

Figure 7. (**a**) compressive strength (UCS) of rocks vs. the ratio *C/A*; (**b**) LA of rocks vs. the ratio *C/A*; (**c**) total porosity (n_t) of rocks vs. the ratio *C/A*.

The results from the investigated sandstone concretes shows that they present satisfactory values of compressive strength (24.00 to 32.00 MPa) relative to other concrete specimens that have been made by andesites and serpentinites as aggregate particles [22]. These satisfactory strength results may attributed to the generally high microtopography of coarse-grained, medium-grained, as well as of fine-grained sandstones relative to the microtopography of other used rocks [22] (Figure 2b,d,f,h and Figure 4). The microtopography of the aggregates constitutes a crucial factor for the mechanical quality of the aggregate rocks and consequently for the quality of the produced concrete as it influences the cohesion and the bonding between the cement paste and the aggregate particles [1,22,23]. The only studied concrete specimen that displays lower strength (25 MPa) than the standard states is the specimen in which the used aggregate was the enriched in carbonate fossils coarse grained sandstone (Figure 2c). This fact may be the result of the lower resistance of the fossils which tend to be broken during the mechanical loading, combined with the low microtopography which they promote (Figure 3d). However, although all the investigated concrete specimens revealed satisfactory strength results, small differences in their values appeared to be dependent on the grain size of the sandstones. The diagrams in Figure 8 indicate that the aggregate properties, which are determined by the size and the number of the grains, as it is shown in Figure 7, determine the final strength of the produced concrete specimens.

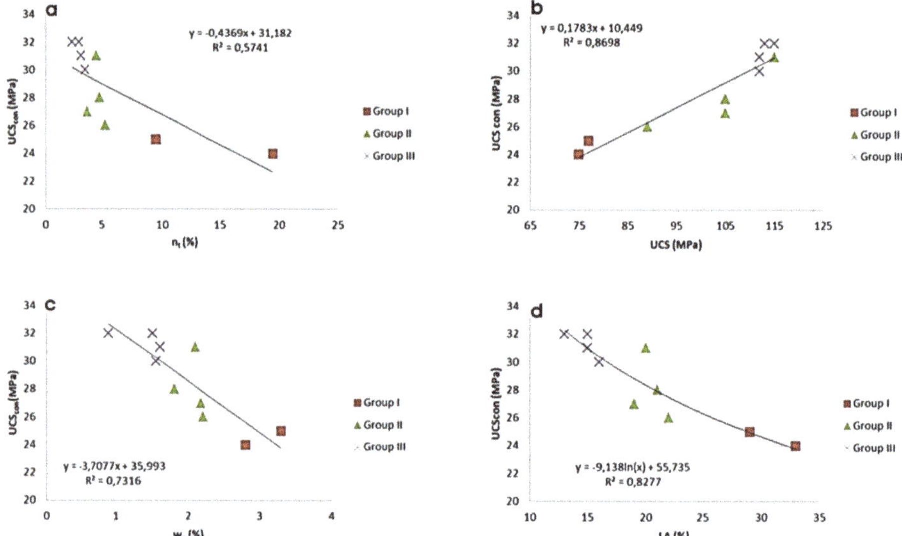

Figure 8. (**a**) Total porosity (n_t) of rocks vs. concrete strength (UCS_{con}); (**b**) compressive strength of rocks vs. concrete strength (UCS_{con}); (**c**) water absorption (w_a) of rocks vs. concrete strength (UCS_{con}); (**d**) LA of rocks vs. concrete strength (UCS_{con}).

Table 4. Correlation equations of diagrams of Figures 6–8.

Diagram	R^2	Equation
6a	0.97	$n_t = 0.658 \times 10^{0.0972LA}$
6b	0.74	$w_a = -0.0381UCS + 5.8771$
6c	0.86	$UCS = -2.2626LA + 147.73$
6d	0.76	$S = 14.776w_a - 7.4628$
7a	0.72	$C/A = 1.1038 \times 10^{0.0317UCS}$
7b	0.90	$C/A = 162.49 \times 10^{-0.087LA}$
7c	0.71	$C/A = -22.81\ln(n_t) + 68.975$
8a	0.57	$UCS_{con} = -0.4369n_t + 31.182$
8b	0.86	$UCS_{con} = 0.1783UCS + 10.449$
8c	0.73	$UCS_{con} = -3.7077w_a + 35.993$
8d	0.82	$UCS_{con} = -9.138\ln(LA) + 55.735$

During the petrographic analysis of the tested concrete specimens, no significant failures and loss of material were observed in those produced by coarse-grained sandstones and nor extensive reaction zones, which typically occur in igneous high porosity aggregates. One possible interpretation that can be attributed is that the lower mechanical strength of concrete aggregates may depend on the higher porosity of the coarse-grained sandstones in contrast to the fine-grained ones (Table 2) which result in the greater adsorption of water which is useful during the 28 days of curing for the achievement of the optimum cohesion between the cement paste and the aggregate particles. However, such extensive areas of incomplete hydration of the cement around the grains were not observed during petrographic examination of the concrete using polarizing microscope. This may have happened because of the even distribution of the mineralogical composites of rocks as can be seen in the 3D depiction via GIS. This resulted in the even adsorption of water and consequently these zones cannot be easily perceived.

5.2. A Potential Scenario for Storage of CO_2 in Sandstones from Klepa Nafpaktias

The studied area presents an appropriate geological basin environment for applying CO_2 capture and storage (CCS) applications. It is well-known that the permeability of flysch formations is regarded as being generally low because of the presence of marl and clay intercalations within this type of formation. This practically impermeable sedimentary formation lies stratigraphically above the sandstones, thus providing an excellent seal caprock to keep the buoyant CO_2 within the reservoir rock. This case presents many similarities with that described for the Mesohellenic Trough (NW Greece), which examined the potential of CO_2 storage within porous sandstones that are overlaid be a less permeable cap rock formation [71–73]. In the latter case, a depth of over 800 m was regarded as suitable for trapping CO_2 under supercritical conditions [73–75]. The sandstone samples provided from our study are highly comparable in terms of composition with sandstones from the Pentalofos formation of the Mesohellenic Trough [72]. Petrographic and mineral modal examinations reveal that the sandstones (Groups I–III) from Klepa Nafpaktias display the following modal compositions: Quartz = 24–29%; K-feldspar = 7–29%; Calcite = 1–8%; Muscovite = 1–4%; Plagioclase~0.5%; Siliceous and Calcite-bearing Cement = 42–57% (Figure 2, Table 1).

These results show that the sandstones examined include relatively higher quartz contents and less calcite compared to those located in the Mesohellenic Trough [73]. Furthermore, effective porosity of the Klepa Nafpaktias sandstones, as it was calculated through the total porosity, which is about 6% for the Group I presents higher values of effective porosity in contrast to the other two sandstone groups and tend to be lower than the Pentalofos sandstones of the Mesohellenic Trough ~ 9%.

Despite the relatively smaller storage potential presented in the region of Klepa Nafpaktias, the rather higher silica contents offers the ability of avoiding undesirable fracture development and disintegration phenomena. This is because CO_2 is expected to react with calcite hosted within the sandstones; however, this would result in the formation of unstable bicarbonates, which would hinder their ability for permanent CO_2 storage. Recent studies on CO_2 geological storage within sandstone formations reveal the importance of feldspar and plagioclase minerals for permanent CO_2 trapping [75–78].

The mineralogical composition of the studied sandstones of Group I as well as their general petrographic characteristics enhances their capacity for CO_2 storage as the sufficient amounts of K-feldspars can react with injected supercritical CO_2 with the following reactions (1)–(4):

$$2KAlSi_3O_8 \text{ (K-feldspar)} + CO_2 + 2H_2O \Rightarrow Al_2(Si_2O_5)(OH)_4 \text{ (kaolinite)} + 4SiO_2 + K_2CO_3 \quad (1)$$

$$3KAlSi_3O_8 \text{ (K-feldspar)} + H_2O + CO_2 \Rightarrow KAl_3Si_3O_{10}(OH)_2 \text{ (illite)} + 6SiO_2 + K_2CO_3 \quad (2)$$

$$2NaAlSi_3O_8 \text{ (albite)} + 2CO_2 + 3H_2O \Rightarrow Al_2(Si_2O_5)(OH)_4 \text{ (kaolinite)} + 4SiO_2 + 2Na^+ + 2HCO^{-3} \quad (3)$$

$$NaAlSi_3O_8 \text{ (albite)} + CO_2 + H_2O \Rightarrow NaAlCO_3(OH)_2 \text{ (dawsonite)} + 3SiO_2 \quad (4)$$

Thus, the dissolution of alkali feldspars will lead to the precipitation of clay minerals and silica (in the form of quartz). Plagioclase, although present in smaller amounts, is also expected to produce kaolinite, as well as calcite through the successive reactions (5) and (6):

$$CaAl_2Si_2O_8 \text{ (anorthite)} + 2CO_2 + 3H_2O \Rightarrow Al_2(Si_2O_5)(OH)_4\text{(kaolinite)} + Ca^{2+} + 2HCO^{-3} \quad (5)$$

$$CaAl_2Si_2O_8 \text{ (anorthite)} + H_2CO_3 + H_2O \Rightarrow Al_2Si_4O_5(OH)_4\text{(kaolinite)} + CaCO_3 \quad (6)$$

We provide preliminary calculations that estimate the CO_2 that could be stored in the frames of a potential pilot project in the studied region. For this aim, we implement this function:

$$CO_2 \text{ Storage Capacity} = \Sigma(V \times \phi \times R\rho \times \varepsilon)$$

With V symbolizing the sandstone reservoir volume (under the flysch cap rock); ϕ denoting the effective porosity; ρ specifying the sCO_2 specific gravity properties; whereas ε stands for the sCO_2 storage ratio capability. We undertake a likely scenario of a test pilot project, in which we assume a volume of 5000 m (the length) × 3000 m (the width) × 500 m (the depth) = 75×10^8. Based upon the estimations of Jin et al. [75] and with reference to the statistical values of USGS modeling, we can consider the CO_2 storage ratio for sandstones to be 1%. The application of this discount factor is necessary in order to obtain a realistic estimation of the sandstone reservoir storage potential. Taking the aforementioned value into consideration, as well as additional parameters that include the average sandstone effective porosity values from our studied site (6%) and the specific gravity considered for CO_2 under supercritical conditions (a value of 400 kg/m^3 in pressure and temperature conditions of 10 MPa and 50 °C respectively [79]), it is assessed that the demarcated area could potentially store an amount of 18×10^5 tons of CO_2.

We also consider Equation (12) of Jin et al. [75] to calculate the quantity of CO_2 trapped by feldspars (K-feldspar and plagioclase minerals, where these amounts K-feldspar = 23–34%; Plagioclase = ~1% resulted from the reduction of the initial amounts without the cement):

$$mCO2 \text{ Feldspar} = [p_{\text{Feldspar}} \times V \times (1 - \varphi) \times X_{\text{Feldspar}} \times M_{\text{CO2}} \times R]/M_{\text{Feldspar}}$$

With V symbolizing the sandstone reservoir volume, ϕ denoting the effective porosity, p_{Feldspar} expresses the feldspar density (2.55–2.67 × 10^3, 2.55–2.60 × 10^3 and 2.75–2.76 × 10^3 kg/m^3 for K-feldspar, albite and anorthite respectively), M_{Feldspar} is molecular weight (279.07, 262.96 and 278.94 for K-feldspar, albite and anorthite respectively) R is the ratio of feldspar mineral to CO_2 0.5, 1 and 1 for K-feldspar, albite and anorthite respectively), X_{Feldspar} the proportions of feldspar minerals, M_{CO2} is the total CO_2 storage capacity of mineral trapping. By applying this equation upon alkali feldspars and plagioclase the results calculated for the CO_2 that can be permanently trapped within the sandstone formation is ~6×10^5 tons, which is less by almost approximately 1/3 of the storage potential calculated with the previous method. This is due to the fact that the latter equation does not consider calcite crystallization as a stable mineral phase. Nevertheless, considering both cases, it is evident that the sandstones of the Klepa Nafpaktias region are capable of storing sufficient amounts of CO_2. This is even more evident taking into consideration that region's sandstones and flysch formations encompasses an even wider area and thus could allow for the deployment of larger-scale CO_2 storage projects, provided that the proposed pilot test is deployed successfully.

6. Conclusions

In this paper, sandstones of various petrographic characteristics derived from Klepa Nafpaktias were examined in order to evaluate their suitability in construction (concrete) and environmental applications (CO_2 storage). For the first time, the petrographic study of rocks such as of those sandstones were carried out by using classic petrographic methods (observation through polarizing microscope) combined with modern tools of quantification of modal composition (GIS proposed method) and 3D depictions of their petrographic features (3D Builder software). The above mentioned study leads to the following concluding remarks:

- Three groups of sandstones were detected according to their petrographic features regarding the grain size (coarse, medium and fine-grained size).
- The above classification of rocks was retained in their physicomechanical and physicochemical properties as well as in the final strength of the produced concrete specimens.
- The petrographic observation of thin sections of the concrete specimens combined with the results of their mechanical strength revealed that the studied sandstones are suitable for concrete aggregates (Groups I–III) except one coarse-grained sample (K.L9 (Group I)) which contains intense amount of carbonate fossils presenting lower concrete strength than the standard states.
- The proposed ratio *C/A* (crystals/mm^2) seems to influence the aggregate properties which constitute critical factors for the final concrete strength, presenting the more fine-grained sandstones as the most suitable for concrete aggregates.
- The petrographic characteristics of the sandstones from Klepa Nafpaktias and their porosity values reveal that the coarse-grained samples (Group I) is more capable for potential CO_2 storage.
- Preliminary calculations suggest that a potential pilot project can store an amount of up to 18×10^5 tons CO_2. The size of the sandstones formation provides the necessary basis for examining the deployment of an even larger scale pilot test that suggested from the present study.

Author Contributions: P.P. was involved in the field data collection, participated in most of the laboratory testing's, result interpretation and paper writing; P.P.G. was involved in the field data collection, participated in most of the laboratory testing's, and contributed to paper writing; A.R. was involved in the field data collection, interpretation, and assisted in the paper writing; M.K. was involved in the field data collection, modified the geological map and performed the GIS analysis; P.K. participated by conducting data interpretation and assisted in part of the paper writing; M.-E.D. was involved in the field data collection and in some of the of the laboratory testing's and N.K. participated by conducting data interpretation and assisted in the paper writing. All authors have read and agreed to the published version of the manuscript.

Funding: Our research study has not received any form funding from external sources.

Acknowledgments: We would like to thank A.K Seferlis for his help concerning the SEM performed work (Lab of Electron Microscopy & Microanalyses, University of Patras).

Conflicts of Interest: No conflicts of interest are declared.

References

1. Petrounias, P.; Giannakopoulou, P.P.; Rogkala, A.; Lampropoulou, P.; Tsikouras, B.; Rigopoulos, I.; Hatzipanagiotou, K. Petrographic and Mechanical Characteristics of Concrete Produced by Different Type of Recycled Materials. *Geosciences* **2019**, *9*, 264. [CrossRef]
2. Farzadnia, N.; Abang, A.A.A.; Demirboga, R.; Anwar, M.P. Effect of halloysite nanoclay on mechanical properties, thermal behavior and microstructure of cement mortars. *Cem. Concr. Res.* **2013**, *48*, 97–104. [CrossRef]
3. Tamanna, N.; Sutan, N.M.; Lee, D.T.C. Utilization of Waste Glass in Concrete. In *Proceedings of the 6th International Engineering Conference, Energy and Environment (ENCON), 1–4 July*; Research Publishing: Wuhan, China, 2013.
4. Castro, S.; Brito, J. Evaluation of the durability of concrete made with crushed glass aggregates. *J. Clean. Prod.* **2013**, *41*, 7–14. [CrossRef]
5. Abdallah, S.; Fan, M. Characteristics of concrete with waste glass as fine aggregate replacement. *J. Eng. Technol. Res.* **2014**, *2*, 11–17.
6. Jani, W.; Hogland, W. Waste glass in the production of cement and concrete—A review. *J. Envir. Chem. Eng.* **2014**, *2*, 1767–1775. [CrossRef]
7. Meng, Y.; Ling, T.G.; Mo, K.H. Recycling of wastes for value-added applications in concrete blocks: An overview. *Resour. Conserv. Recycl.* **2018**, *138*, 298–312. [CrossRef]

8. Poon, C.S.; Chan, D. Paving blocks made with recycled concrete aggregate and crushed clay brick. *Constr. Build. Mater.* **2006**, *20*, 569–577. [CrossRef]
9. Vanitha, S.; Natrajan, V.; Prada, M. Utilization of waste plastics as a partial replacement of coarse aggregate in concrete blocks. *Indian J. Sci. Technol.* **2015**, *8*, 256–268. [CrossRef]
10. Jackson, N. *Civil Engineering Materials*; Macmillan Press Ltd.: London, UK, 1981.
11. LaLonde, W.S.; Janes, M.F. *Concrete Engineering Handbook*; Library of Congress: New York, NY, USA, 1961.
12. *US Concrete Industry Report*; Library of Congress: New York, NY, USA, 2001.
13. Neville, A.M. *Properties of Concrete, ELSB*, 5th ed.; Pearson Education Publishing Ltd.: London, UK, 2005.
14. Taylor, G.D. *Materials in Construction*, 2nd ed.; Longman Group Ltd., Longman House, Burnt Mill: Harlow, UK, 1994.
15. Neville, A.M. *Properties of Concrete*, 4th ed.; Pitman: London, UK, 1995.
16. Al-Oraimi, S.K.; Taha, R.; Hassan, H.F. The effect of the mineralogy of coarse aggregate on the mechanical properties of high-strength concrete. *Constr. Build. Mater.* **2006**, *20*, 499–503. [CrossRef]
17. Mackechnie, J.R. Shrinkage of concrete containing greywake sandstone aggregate. *ACI Mater. J.* **2006**, *103*, 390–396.
18. Rodgers, M.; Hayes, G.; Healy, M.G. Cyclic loading tests on sandstone and limestone shale aggregates used in unbound forest roads. *Constr. Build. Mater.* **2009**, *23*, 2421–2427. [CrossRef]
19. Verstrynge, E.; Schueremans, L.; Van Gement, D. Creep and failure prediction of diestian ferruginous sandstone: Modelling and repair options. *Constr. Build. Mater.* **2012**, *29*, 149–157. [CrossRef]
20. Kumar, S.; Gupta, R.C.; Shrivastava, S. Strength, abrasion and permeability studies on cement concrete containing sandstone coarse aggregates. *Constr. Build. Mater.* **2016**, *125*, 884–891. [CrossRef]
21. Yilmaz, M.; Tugrul, A. The effects of different sandstone aggregates on concrete strength. *Constr. Build. Mater.* **2012**, *35*, 294–303. [CrossRef]
22. Petrounias, P.; Giannakopoulou, P.P.; Rogkala, A.; Stamatis, P.M.; Tsikouras, B.; Papoulis, D.; Lampropoulou, P.; Hatzipanagiotou, K. The Influence of Alteration of Aggregates on the Quality of the Concrete: A Case Study from Serpentinites and Andesites from Central Macedonia (North Greece). *Geosciences* **2018**, *8*, 115. [CrossRef]
23. Petrounias, P.; Giannakopoulou, P.P.; Rogkala, A.; Stamatis, P.M.; Lampropoulou, P.; Tsikouras, B.; Hatzipanagiotou, K. The Effect of Petrographic Characteristics and Physicomechanical Properties of Aggregates on the Quality of Concrete. *Minerals* **2018**, *8*, 577. [CrossRef]
24. Li, Y.; Onasch, M.C.; Guo, Y. GIS-based detection of grain boundaries. *J. Struct. Geol.* **2008**, *30*, 431–443. [CrossRef]
25. Barraud, J. The use of watershed segmentation and GIS software for textural analysis of thin sections. *J. Volcanol. Geotherm. Res.* **2006**, *154*, 17–33. [CrossRef]
26. Fernandez, F.J.; Menendez-Duarte, R.; Aller, J.; Bastida, F. Application of geographical information systems to shape-fabric analysis. In *High-Strain Zones: Structure and Physical Properties, 245*; Bruhn, D., Burlini, L., Eds.; Geological Society of London Special Publication: London, UK, 2005; pp. 409–420.
27. Tarquini, S.; Favalli, M. A Microscopic Information System (MIS) to assist in petrographic analysis. *Comput. Geosci.* **2010**, *36*, 665–674. [CrossRef]
28. Becattini, V.; Motmans, T.; Zappone, A.; Madonna, C.; Haselbacher, A.; Steinfeld, A. Experimental investigation of the thermal and mechanical stability of rocks for high-temperature thermal-energy storage. *Appl. Energy* **2017**, *203*, 373–389. [CrossRef]
29. Kuravi, S.; Trahan, J.; Goswami, D.Y.; Rahman, M.M.; Stefanakos, E.K. Thermal energy storage technologies and systems for concentrating solar power plants. *Prog. Energy Combust. Sci.* **2013**, *39*, 285–319. [CrossRef]
30. Khare, S.; Dell' Amico, M.; Knight, C.; Mc Garry, S. Selection of materials for hightemperature sensible energy storage. *Solar Energy Mater. Solar Cells* **2013**, *115*, 114–122. [CrossRef]
31. Allen, K.G.; von Backström, T.W.; Kröger, D.G.; Kisters, A.F.M. Rock bed storage for solar thermal power plants: Rock characteristics, suitability, and availability. *Solar Energy Mater. Solar Cells* **2014**, *126*, 170–183. [CrossRef]
32. Tiskatine, R.; Eddemani, A.; Gourdo, L.; Abnay, B.; Ihlal, A.; Aharoune, A.; Bouirden, L. Experimental evaluation of thermo-mechanical performances of candidate rocks for use in high temperature thermal storage. *Appl. Energy* **2016**, *171*, 243–255. [CrossRef]

33. Karakitsios, V.; Tzortzaki, E.; Giraud, F.; Pasadakis, N. First evidence for the early Aptian Oceanic Anoxic Event (OAE1a) from the Western margin of the Pindos Ocean (NW Greece). *Geobios* **2018**, *51*, 187–210. [CrossRef]
34. Robertson, A.H.F.; Karamata, S. The role of subduction-accretion processes in the tectonic evolution of the Mesozoic Tethys in Serbia. *Tectonophysics* **1994**, *234*, 73–94. [CrossRef]
35. Aubouin, J.; Bonneau, M.; Davidson, G.J.; Leboulenger, P.; Matesko, S.; Zambetakis, A. Esquisse structurale de l'Arc egeen externe: Des Dinarides aux Taurides. *Bull. Soc. Géol. Fr.* **1976**, *7*, 327–336. [CrossRef]
36. Bernoulli, D.; De Graciansky, P.C.D.; Monod, O. The extension of the Lycian Nappes (SW Turkey) into the Southeastern Aegean Islands. *Eclogae Geol. Helv.* **1974**, *67*, 39–90.
37. Argyriadis, I.; De Graciansky, P.C.; Marcoux, J.; Ricou, L.E. The opening of the Mesozoic Tethys between Eurasia and Arabia-Africa. In Proceedings of the Geologie des Chaınes Alpines Issues de la Tethys, 26th International Geological Congress, Colloque C5, Paris, France, 7–17 July 1980; Aubouin, J., Debelmas, J., Latreille, M., Eds.; Bureau de Recherches Geologiques et Minieres Memoire: Paris, France, 1980; Volume 115, pp. 199–214.
38. Kafousia, N.; Karakitsios, V.; Jenkyns, H.C.; Mattiolis, E. A global event with a regional character: The Early Toarcian Oceanic Anoxic Event in the Pindos Ocean (northern Peloponnese, Greece). *Geol. Mag.* **2011**, *148*, 619–631. [CrossRef]
39. Fleury, J.J. Les zones de Gavrovo-Tripolitza et du Pinde-Olonos (Grece continentale et Peloponese du nord) Evolution d'une plate-forme et d'un basin dans leur cadre alpin. *Soc. Geol. Nord* **1980**, *4*, 1–473.
40. De Wever, P.; Baudin, F. Palaeogeography of radiolarite and organic-rich deposits in Mesozoic Tethys. *Geol. Rundsch.* **1996**, *85*, 310–326. [CrossRef]
41. Clift, P.D. The collision tectonics of the southern Greek Neotethys. *Geol. Rundsch.* **1992**, *81*, 669–679. [CrossRef]
42. Degnan, P.J.; Robertson, A.H.F. Mesozoic–early Tertiary passive margin evolution of the Pindos Ocean (NW Peloponnese Greece). *Sediment. Geol.* **1998**, *117*, 33–70. [CrossRef]
43. Pe-Piper, G. The nature of Triassic extension-related magmatism in Greece: Evidence from Nd and Pb isotope geochemistry. *Geol. Mag.* **1998**, *135*, 331–348. [CrossRef]
44. Neumann, P.; Zacher, W. The Cretaceous sedimentary history of the Pindos Basin Greece. *Int. J. Earth Sci.* **2004**, *93*, 119–131. [CrossRef]
45. Jones, G.; Robertson, A.H.F.; Cann, J.R. Genesis and emplacement of the suprasubduction zone Pindos Ophiolite, Northwestern Greece. In *Ophiolite Genesis and Evolution of the Oceanic Lithosphere*; Peters, T., Nicolas, A., Coleman, S., Eds.; Sultanate of Oman Ministry of Petroleum and Minerals: Muscat, Oman, 1991.
46. Konstantopoulos, P.A.; Zelilidis, A. Sedimentation of submarine fan deposits in the Pindos foreland basin, from late Eocene to early Oligocene, west Peloponnesus peninsula, SW Greece. *Geol. J.* **2013**, *48*, 335–362. [CrossRef]
47. Faupl, P.; Pavlopoulos, A.; Migiros, G. On the provenance of flysch deposits in the External Hellenides of mainland Greece: Results from heavy mineral studies. *Geol. Mag.* **1999**, *135*, 412–442. [CrossRef]
48. Vakalas, I. Evolution of Foreland Basins in Western Greece. Ph.D. Thesis, University of Patras, Patras, Greece, 2004.
49. Loftus, D.L.; Matarangas, D.; Zindros, G.; Katsikatsos, G. *Geological Map of Greece, Klepa Sheet, 1:50,000*; IGME: Athens, Greece, 1984.
50. *Part 1: Composition, Specifications and Conformity Criteria for Common Cements*; EN 197-1; European Standard: Warsaw, Poland, 2011.
51. *Part 3: Procedure and Terminology for Simplified Petrographic Description*; EN 932; European Standard: Warsaw, Poland, 1996.
52. *Methods for Sampling and Testing of Mineral Aggregates, Sands and Fillers, Part 1: Methods for Determination of Particle Size and Shape*; BS 812; British Standard Institution: London, UK, 1975.
53. ISRM Suggested Methods. *Rock Characterization Testing and Monitoring*; Brown, E., Ed.; Pergamon Press: Oxford, UK, 1981.
54. EN 1367-2. *Tests for Thermal and Weathering Properties of Aggregates—Part 2: Magnesium Sulfate Test*; European Committee for Standardization: Brussels, Belgium, 1999.
55. EN 1097-6. *Tests for Mechanical and Physical Properties of Aggregates—Part 6: Determination of Particle Density and Water Absorption*; European Committee for Standardization: Brussels, Belgium, 2000.

56. ASTM C-131. *Resistance to Abrasion of Small-Size Coarse Aggregate by Use of the Los Angeles Machine*; American Society for Testing and Materials: Philadelphia, PA, USA, 1989.
57. *Standard Test Method for Unconfined Compressive Strength of Intact Rock Core Specimens*; ASTM D 2938-95; American Society for Testing and Materials: West Conshohocken, PA, USA, 2002.
58. *Standard for Selecting Proportions for Normal, Heavyweight and Mass Concrete*; ACI-211.1-91; American Concrete Institute: Farmington Hills, MI, USA, 2002.
59. *Part 3: Testing Hardened Concrete. Compressive Strength of Test Specimens*; British Standard Institution: London, UK, 2009; BS EN 12390.
60. *Standard Practice for Petrographic Examination of Hardened Concrete*; ASTM C856; American Society for Testing and Materials: West Conshohocken, PA, USA, 2017.
61. Rigopoulos, I.; Tsikouras, B.; Pomonis, P.; Hatzipanagiotou, K. The impact of petrographic characteristics on the engineering properties of ultrabasic rocks from northern and central Greece. *Q. J. Eng. Geol. Hydrogeol.* **2012**, *45*, 423–433. [CrossRef]
62. Smith, M.R.; Collis, L. *Aggregates: Sand, Gravel and Crushed Rock Aggregates for Construction Purposes*; Spec. Publ. 17; The Geological Society: London, UK, 2001.
63. Hartley, A. A review of the geological factors influencing the mechanical properties of road surface aggregates. *Q. J. Eng. Geol.* **1974**, *7*, 69–100. [CrossRef]
64. Barttli, B. The influence of geological factors on the mechanical properties of basic igneous rocks used as road surface aggregates. *Eng. Geol.* **1992**, *33*, 31–44. [CrossRef]
65. Petrounias, P.; Giannakopoulou, P.P.; Rogkala, A.; Lampropoulou, P.; Koutsopoulou, E.; Papoulis, D.; Tsikouras, B.; Hatzipanagiotou, K. The Impact of Secondary Phyllosilicate Minerals on the Engineering Properties of Various Igneous Aggregates from Greece. *Minerals* **2018**, *8*, 329. [CrossRef]
66. Turgul, A.; Zarif, I.H. Correlation of mineralogical and textural characteristics with engineering properties of selected granitic rocks from Turkey. *Eng. Geol.* **1999**, *51*, 303–317.
67. Rigopoulos, I.; Tsikouras, B.; Pomonis, P.; Hatzipanagiotou, K. Correlations between petrographic and geometrical properties of ophiolitic aggregates from Greece. *Bull. Eng. Geol. Environ.* **2014**, *73*, 1–12. [CrossRef]
68. Escartin, J.; Hirth, G.; Evans, B. Strength of slightly serpentinized peridotites: Implications for the tectonics of oceanic lithosphere. *Geology* **2001**, *29*, 1023–1026. [CrossRef]
69. Giannakopoulou, P.P.; Petrounias, P.; Rogkala, A.; Tsikouras, B.; Stamatis, P.M.; Pomonis, P.; Hatzipanagiotou, K. The influence of the mineralogical composition of ultramafic rocks on their engineering performance: A case study from the Veria-Naousa and Gerania ophiolite complexes (Greece). *Geosciences* **2018**, *8*, 251. [CrossRef]
70. Giannakopoulou, P.P.; Petrounias, P.; Tsikouras, B.; Kalaitzidis, S.; Rogkala, A.; Hatzipanagiotou, K.; Tombros, S.F. Using Factor Analysis to Determine the Interrelationships between the Engineering Properties of Aggregates from Igneous Rocks in Greece. *Minerals* **2018**, *8*, 580. [CrossRef]
71. Koukouzas, N.; Ziogou, F.; Gemeni, V. Preliminary assessment of CO_2 geological storage opportunities in Greece. *Int. J. Greenh. Gas Con.* **2009**, *3*, 502–513. [CrossRef]
72. Tassianas, A.; Koukouzas, N. CO_2 Storage Capacity Estimate in the Lithology of the Mesohellenic Trough, Greece. *Energy Procedia* **2016**, *86*, 334–341. [CrossRef]
73. Koukouzas, N.; Kypritidou, Z.; Purser, G.; Rochelle, C.A.; Vasilatos, C.; Tsoukalas, N. Assessment of the impact of CO_2 storage in sandstone formations by experimental studies and geochemical modeling: The case of the Mesohellenic Trough, NW Greece. *Int. J. Greenh. Gas Con.* **2018**, *71*, 116–132. [CrossRef]
74. Shafeen, A.; Croiset, E.; Douglas, P.L.; Chatzis, I. CO_2 sequestration in Ontario, Canada. Part I: Storage evaluation of potential reservoirs. *Energy Convers. Manag.* **2004**, *45*, 2645–2659. [CrossRef]
75. Jin, C.; Liu, L.; Li, Y.; Zeng, R. Capacity assessment of CO_2 storage in deep saline aquifers by mineral trapping and the implications for Songliao Basin, Northeast China. *Energy Sci. Eng.* **2017**, *5*, 81–89. [CrossRef]
76. Ryoji, S.; Thomas, L.D. Experiment al study on water-rock interactions during CO_2 flooding in the Tensleep Formation, Wyoming, USA. *Appl. Geochem.* **2000**, *15*, 265–279.
77. Robert, J.R.; Tamer, K.; James, L.P. Experiment al investigation of CO_2- brine-rock interactions at elevated temperature and pressure: Implications for CO_2 sequestration in deep-saline aquifers. *Fuel Process. Technol.* **2005**, *86*, 1581–1597.

78. Ryzhenko, B.N. Genesis of dawsonite mineralization: Thermo-dynamic analysis and alt ernative. *Geochem. Int.* **2006**, *44*, 835–840. [CrossRef]
79. Spycher, N.; Pruess, K. CO_2-H_2O Mixtures in the Geological Sequestration of CO_2. II. Partitioning in Chloride Brines at 12–100 °C and up to 600 bar. *Geochim. Cosmochim. Acta* **2005**, *69*, 3309–3320. [CrossRef]

Article

Acid-Catalyzed Wet Torrefaction for Enhancing the Heating Value of Barley Straw

Antonios Nazos [1], Panagiotis Grammelis [2,*], Elias Sakellis [3] and Dimitrios Sidiras [1]

[1] Laboratory of Simulation of Industrial Processes, Department of Industrial Management and Technology, School of Maritime and Industrial Studies, University of Piraeus, 80 Karaoli & Dimitriou, GR 18534 Piraeus, Greece; anazos@yahoo.gr (A.N.); sidiras@unipi.gr (D.S.)

[2] Technology Hellas/Chemical Process and Energy Resources Institute (CERTH/CPERI), Centre for Research, Athens branch: Egialias 52, GR-15125 Marousi, Athens, Greece

[3] Institute of Nanoscience and Nanotechnology, National Centre for Scientific Research "Demokritos", GR-15310 Agia Paraskevi Attikis, Athens, Greece; e.sakellis@inn.demokritos.gr

* Correspondence: grammelis@certh.gr; Tel.: +30-211-106-9504

Received: 25 February 2020; Accepted: 24 March 2020; Published: 3 April 2020

Abstract: In the present study, the possibility of improving the higher heating value (HHV) of lignocellulosic biomass, especially barley straw, was examined. The research deals with the treatment of barley straw by acid-catalyzed wet torrefaction (ACWT), also called acid hydrolysis, in a batch reactor (autoclave) Parr 4553 3.75 L. In this case, two different simulation approaches were applied: (i) combined severity factor (CSF) and (ii) response surface methodology (RSM) based on Box–Behnken design of experiments (DoE). Sulfuric acid (SA) concentration, temperature and time were the ACWT parameters examined herein. An oxygen bomb calorimeter was used for the HHV measurement. The findings indicated that the composition changes of the straw due to ACWT had a significant effect on the HHV of the pretreated material. In this study, treatment conditions were 10–35 mM SA, 160–200 °C and an isothermal reaction time 0–40 min (preheating period not included in these values). In conclusion, there was a significant increase in the HHV up to 24.3 MJ/kg for the ACWT barley straw, compared to 17.5 MJ/kg for the untreated straw, at optimal conditions of 200 °C for 25 min (isothermal period) and 35 mM SA. This resulted in a 1.39 enhancement factor (EF) and 68% energy yield (EY).

Keywords: acid-catalyzed wet torrefaction; acid hydrolysis; barley straw; combined severity factor; enhancement factor; energy yield; higher heating value; response surface methodology

---

1. Introduction

The increase in worldwide energy demand has a significant effect on the fossil fuel contribution to environmental pollution and climate change, provoking a global interest in the use of environmentally friendly, renewable fuels [1,2]. Renewable biomass is one of the emerging energy resources with high potentials that can balance CO_2 emissions [3]. Lignocellulosic biomass is currently a major energy source for fossil fuel partial substitution [4–6]. Biomass contributes about 10% of the global annual energy production [7,8]. Thus, biomass attracts considerable research interest in order to meet the increasing future sustainable energy demand [9]. Biomass is widely available in nature and accounts for around 100 billion tons per year [6]. Heat, electricity, fuel, chemicals and other high added value products can be produced from biomass [7,9].

Moreover, biomass disadvantages include heterogeneity in structure, low density, high humidity, low higher heating values (HHVs) and flexibility, all of which limit the use of biomass as a fuel [7,10]. These disadvantages make the production of heat and energy from biomass more complicated. They result in the increased cost of processing, transporting and storing of biomass fuels. Consequently, we need to process the fuel and upgrade it through processes like the torrefaction process. Combined with understanding and studying the mechanisms of the biomass reaction through the torrefaction process, it will be easier to develop the technologies for thermal utilization of biomass [7,11–15].

There are mainly two torrefaction categories in the recent literature: dry torrefaction (DT) and wet torrefaction (WT). Both these techniques produce biomass fuels with improved chemical and physical qualities compared to the untreated biomass characteristics [7]. No references were found with regard to acid-catalyzed wet torrefaction (ACWT) but it could be considered as a subcategory of WT which uses, e.g., SA as catalyst (not autocatalyzed by organic acids produced during WT). On the other hand, acid hydrolysis was examined for production of fermentable to bioethanol sugars from lignocellulosic biomass [16,17]. However, DT has been extensively studied for decades and significant references appear in the international literature [13–15] while WT appeared later, and is associated with concepts such as autohydrolysis, hydrothermal carbonization, hydrothermal pretreatment, etc. In recent years, research on WT has increased compared to investigation on DT [7,18,19].

WT is the treatment of biomass with hydrothermal media at 180–260 °C with pressures at 0.9–4.6 MPa, respectively [4,7,20,21]. WT seems to be a promising technology to produce low cost high quality solid biomass fuels from agricultural wastes. WT is synonymous with "hydrothermal carbonization" or "hydrothermal conversion" [22] or "hydrothermal treatment" [23]. Sometimes WT and hydrothermal carbonization terminologies are used alternatively. Although WT is applied to produce advanced solid fuels, hydrothermal carbonization is primarily used for coal production. WT gives fuels with an increased HHV, as well as low carbon emissions, which can also be considered as moderate biomass pyrolysis at 200–300 °C [24]. WT produces solid fuels with enhanced properties at comparatively milder conditions with regard to time and temperature [25]. Furthermore, WT was considered to be a low cost and thus cost-effective prefabrication method with promising equipment corrosion-limiting properties and simple operation [4,7]. After All, WT or hydrothermal pretreatment has attracted an interest, being an effective way to convert high moisture biomass [26]. Recently, Gan et al. [27] developed microwave-assisted ACWT (using various acids as catalysts) of microalgae for simultaneous production of char and sugars.

According to He et al. [4], issues associated with operating cost, pollutant emissions, re-design of industrially scale reactor, and system integration with downstream applications must be solved in order to make WT environmentally and commercially sustainable. The most critical issues deal with (i) requirements of reactor materials withstanding high temperature, moderate pressure and severe corrosion; (ii) clogging issues caused by inorganic precipitates; (iii) post-treatment of wastewater and downstream application of WT material; and (iv) heat recovery to minimize energy consumption.

Furthermore, in ACWT the recovery and recycling of the used acid (e.g., sulfuric acid) is crucial. According to Gan et al. [27] the corrosive nature of sulfuric acid requires a high capital cost for the pretreatment reactor, as well as high safety and handling measures due to using acid. Moreover, the generated byproducts inhibit bioethanol production from the sugars in the ACWT liquid phase and requires costly downstream waste treatment.

In our recently published work [28], the effect of DT in a muffle furnace on barley straw HHV was simulated using the combined severity factor (CSF) methodology (incorporating the effect of reaction time and temperature). In this study, in an effort to optimize the HHV of barley straw, the feasibility of ACWT was examined. Thus, the subject of this research is the modification of barley straw by acid hydrolysis using a batch reactor (autoclave). The pretreatment conditions were time, temperature and acid concentration. Moreover, two approaches were used to simulate the HHV enhancement during the performed experiments. These approaches were (i) the response surface methodology (RSM) based on a Box–Behnken design of experiments (DoE) and (ii) CSF. The HHV of the ACWT barley straw was compared to that of the untreated material. Afterwards, the ACWT conditions were optimized for maximizing the HHV of modified barley straw.

2. Materials and Methods

2.1. Material Development

The experiments performed using barley straw, originated in the Kapareli village of Thebes, Greece (38°14′8″ N 23°12′59″ E). The original straw was manually cut to particles of 2 and 3 cm. The specific fraction was considered to be suitable because in this way homogeneity could be achieved when the ACWT procedure was over. The untreated straw moisture was 6.0% *w*/*w* measured according to the procedure UNE-EN ISO 18134-1: 2015. Moreover, the barley straw used for the experiments in this work was the same raw material used in an earlier study of our team on DT [28].

2.2. ACWT Process

To carry out the ACWT treatment, a Parr 4553 batch reactor 3.75 L was used, being capable of reaching working temperatures up to 250 °C, while the planned experiments require a constant prevailing temperature of 160 or 180 or 200 °C. For each temperature level, there was a set of experiments which combine different acid concentration and reaction time. More specifically, the different SA concentrations tested herein were 0.01, 0.0225 and 0.035 M over 0, 20 or 40 min of reaction time each (isothermal reaction periods, not including the preheating and cooling period). When the sample has remained in the reactor vessel for the planned time, the cooling process kicks in, cooling the vessel to approximately 25 °C. It should be noted that throughout each experiment, the liquid/solid solution was stirred at 150 rpm, while the vessel preheating and the duration of the cooling process is neglected. The ACWT process was followed by a separation and drying process. In particular, the samples were extracted from the liquid solution using a No. 3 Whatman filter, washed and finally dried at 105 °C using an oven. The solid residue yield (SRY) was estimated by comparing the substrate weight, before and after the ACWT process, i.e., as the fraction substrate used for torrefaction.

2.3. Bomb Calorimeter

A Parr 1341 Plain Jacket Bomb Calorimeter was used for the HHV measurements. The original barley straw and modified barley straw samples produced by ACWT were tested for HHV determination according to the method ISO 1928:2009 "Solid mineral fuels—Determination of gross calorific value by the bomb calorimetric method and calculation of net calorific value". More details are given in a recent work of our team [28].

2.4. Proximate and Ultimate Analysis

The moisture content, the volatile matter (VM), the fixed carbon (FC), the ash, the carbon, the hydrogen, the nitrogen, the sulfur and the oxygen (by difference) of untreated and ACWT pretreated samples was determined by proximate and ultimate analysis.

2.5. Scanning Electron Microscopy

Scanning electron microscopy (SEM) was used to examine the surface topology changes of the untreated barley straw compared to the ACWT treated barley straw. For this purpose, a FEI INSPECT SEM equipped with an EDAX super ultra-thin window analyzer for energy-dispersive X-ray spectroscopy (EDS) was used.

2.6. Combined Severity Factor

In order to incorporate the effect of reaction conditions (SA concentration, time and temperature) into a single ordinate we applied the CSF. The CSF concept was based on the "*P* factor" or "reaction ordinate" presented by Brasch and Free [29] in 1965 and Overend and Chornet [30] in 1987:

$$P \text{ factor} = [\exp((T - 100)/14.75] \cdot t \quad (1)$$

where T represents the temperature and t is the time of the procedure.

In 1990, Chum et al. [31], and in 1992, Abatzoglou et al. [32] expanded this abovementioned formula, including the effects induced by how acidic or basic the solution is, according to the following equation:

$$R_0' = 10^{-\mathrm{pH}} \cdot t \cdot \exp[(T - 100)/14.75] \tag{2}$$

The CSF concept enables the comparison and\or unifications of different pretreatments. CSF has been widely used. For instance, it was used on acid hydrolysis (used as a first step before enzymatic saccharification) for softwood and corn stover samples, in the study Lloyd and Wyman [33], as well as for samples of wheat straw by Kabel et al. [34]. However, the latter study is the only one characterized by low CSF, to the best of our knowledge. In particular, CSF in the study of Kabel et al. ranges from −1.7 to 1.5, while in contrast Lloyd and Wyman's CSF ranges from 0.4 to 2.7.

CSF also can be used for non-isothermal acid treatment, based on previous studies [35–37], with the major difference from Equation (2) laying in expressing with more detail the change of temperature over time (including preheating and cooling period over the commonly used isothermal period):

$$R_0^* = e^{-\mathrm{pH}} \int_0^t \exp\left(\frac{(T - 100)}{14.45}\right) dt \tag{3}$$

The pH was measured in the liquid phase derived after ACWT (i.e., acid hydrolysis pretreatment). In the present work, the CSF is R_0^* as expressed by Equation (3).

2.7. Statistical Method

As a second method (additional to CSF), the RSM is used to incorporate the effect of the ACWT variables (temperature, residence time and SA concentration) into the modified barley straw HHV. The RSM was based on the Box–Behnken DoE (BBD). This particular design is considered more efficient than other designs, such as central composite design (CCD) and three-level full factorial design [38]. The CCD is a five-level fractional factorial design, which comprises of a two-level factorial design, central designs and two axial designs. On the other hand, the BBD is a spherical, rotatable second-order design. It is based on a three-level incomplete factorial design, which consists of the center point and middle points like the edge of a cube. Although BBD can be derived from a cube, it can be represented spherically, making the vertices of the cube not covered by the design. It can be considered as three interlocking factorial designs along with center points. The BBD is said to be a more economical and viable tool than the CCD, because its design matrix is usually generated with a fewer number of experimental runs. BBD requires an experiment number according to $N = 2k(k - 1) + c_p$, where k is the number of factors and c_p is the number of the central points, while CCD requires $N = k^2 + 2k + c_p$. More specifically, CCD requires experiment number $N = 18$ while BBD requires only $N = 15$ for $k = 3$ and $c_p = 3$.

The BBD method is advantageous over other common experimental or optimization methods since it requires a relatively small number of experiments while it also enables the exploration of the interactive effects over the considered observables, hence enabling to depict the effects of each one [38].

The experiments are planned with Quantum XL (SigmaZone) software and are presented in Table 1. It should be noted that each of the 15 experiments was performed twice, with Table 1 containing the averaged outcomes.

Table 1. Box–Behnken experimental design.

Run	T (°C)	t (min)	SA (mol/m^3)	pH before	pH after
1	160	0	22.50	1.74	2.37
2	200	0	22.50	1.74	2.05
3	160	40	22.50	1.74	2.30
4	200	40	22.50	1.74	2.40
5	180	0	10.00	3.31	4.02
6	180	0	35.00	1.75	1.80
7	180	40	10.00	2.49	3.64
8	180	40	35.00	1.79	1.90
9	160	20	10.00	2.8	3.97
10	160	20	35.00	1.79	1.93
11	200	20	10.00	3.31	3.53
12	200	20	35.00	1.73	1.89
13	180	20	22.50	1.88	2.60
14	180	20	22.50	2.14	2.63
15	180	20	22.50	2.14	2.65

Performing a polynomial regression with the method of least squares as described from Box and Wilson [39], enables the prediction of optimum conditions through RSM. More accurately, in order to draw a three-dimensional graph (the RSM) that will facilitate the identification of the optimum data, the most important steps include the determination of influencing variables and their range, the selection of a random point on sample as representative of input, encoding of the variables, implementation of the analysis of variance (ANOVA) technique and the lack-of-fit test.

The experiments outcome (*A*: temperature, *B*: time and *C*: SA concentration) fitted to the following equation:

$$\mathrm{HHV} = a + a_1A + a_2B + a_3C + a_{11}A^2 + a_{22}B^2 + a_{33}C^2 + a_{12}AB + a_{13}AC + a_{23}BC \tag{4}$$

where a, a_{xx} are the model's constant and coefficients, respectively.

3. Results and Discussion

3.1. Simulation Based on CSF

The pH before and after ACWT of barley straw depending mostly on the concentration of SA is given in Table 1. Furthermore, the HHV and the SRY of barley straw after ACWT are presented in Table 2. According to the results, ACWT had a mild effect on barley straw SRY. Moreover, in Table 2, the HHV as affected by the ACWT treatment conditions is presented. These findings indicate that there was a serious increase in HHV for high temperature and the maximum SA concentration while untreated barley straw HHV was measured and its average rate was 17.6 MJ/Kg. This was the average result of two HHV measurements.

The CSF of the present study incorporates the effect of temperature, time and SA concentration. In Figure 1a the reaction temperature profile vs. the reaction time is presented. The isothermal period was 40 min at 160, 180 and 200 °C. In Figure 1b the reaction pressure profile vs. time for the same experiments is presented. Table 2 shows the CSF values estimated for the specific ACWT conditions. The CSF in logarithmic form ranged from −0.70 to 2.61, incorporating the effect of temperature, time and SA concentration. Combined increase in temperature, time and SA concentration leads to higher values of the CSF. Relevant research on straw pretreatment [40] shows that severe hydrolytic conditions leads to the degradation of cellulose and hemicelluloses, breaks the lignocellulosic matrix, decreases the ash content and changes the elemental composition of the modified material. In another relevant work, Angles et al. [41] used the CSF (also called gravity factor) for the softwood vapor explosion at 176 to 231 °C for 2.5 to 5.5 min hydrolysis. In their work the CSF varied from $\log R_0^* = 2.6$

to $\log R_0^* = 4.6$. Moreover, in the Chornet and Overend [30] study CSF was 2.6–4.2. Decreased sugar yield and degree of polymerization for severe hydrolytic conditions were reported. Toussaint et al. [42] and Heitz et al. [43] proved that a higher reaction temperature and time (i.e., increased CSF) results in enhanced cellulose recovery, accessibility and digestibility in combination with higher lignin removal. This was followed by increased hemicellulose degradation.

Table 2. Combined severity factor (CSF) in logarithmic form ($\log R_0^*$), higher heating value (HHV) and solid residue yield (SRY) of barley straw treated by acid-catalyzed wet torrefaction (ACWT).

Run	T (°C)	t (min)	SA (mol/m^3)	$\log R_0^*$	SRY % wt.	HHV (MJ/kg)
1	160	0	22.5	0.36	60.50	19.17
2	200	0	22.5	1.89	47.40	20.10
3	160	40	22.5	1.22	50.30	20.13
4	200	40	22.5	2.30	44.70	22.17
5	180	0	10.0	−0.70	59.20	19.53
6	180	0	35.0	1.55	47.40	19.71
7	180	40	10.0	0.47	48.00	20.47
8	180	40	35.0	2.21	49.50	20.42
9	160	20	10.0	−0.64	54.00	19.43
10	160	20	35.0	1.42	56.60	21.78
11	200	20	10.0	0.97	47.50	23.44
12	200	20	35.0	2.61	31.60	24.27
13	180	20	22.5	1.31	47.00	20.48
14	180	20	22.5	1.28	50.10	20.28
15	180	20	22.5	1.28	48.20	20.43

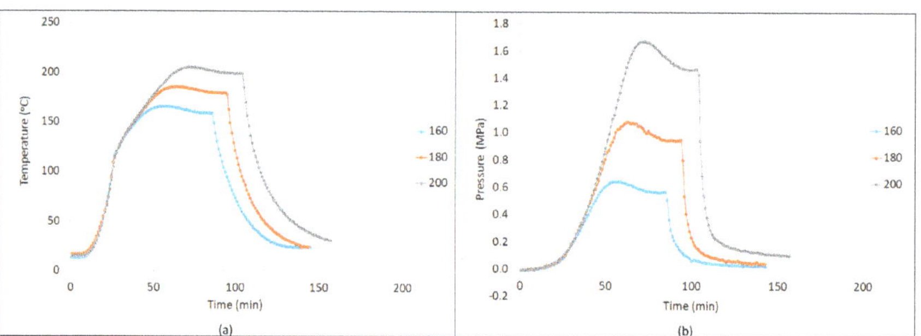

Figure 1. The ACWT pretreatment's temperature (**a**) and pressure (**b**) profiles vs. time (isothermal period 40 min at 160, 180 and 200 °C, respectively).

Demirbas [44] reported that the HHV of a lignocellulosic fuel depends on lignin, cellulose and hemicellulose percentage. He found that holocellulose HHV was 18.60 MJ/kg, while lignin HHV was 23.26–25.58 MJ/kg. Consequently, the HHV of lignocellulosic fuels is improved when the lignin content is increased.

Changes in the elemental composition of lignocellulosics due to ACWT affects the HHV of the solid residue. According to Semhaoui et. al. [45] hemicelluloses solubilization depends on the severity of treatment and becomes significant for positive CSF values. The results of CSF show that the concentration of acid that led to negative levels CSF value was insufficient for hemicelluloses solubilization.

In the present work, the correlation between $\log R_0^*$ and SRY % w/w (dry basis) was found by applying a nonlinear regression, fitting the equation

$$\text{SRY} = 3.857\exp(-1.480x) + 47.92 \quad (5)$$

where $x = \log R_0^*$, and the standard error of the estimate (SEE) was equal to 2.885. In Figure 2, the SRY percentage decreases significantly when the CSF increases. This decrease becomes negligible for relatively high $\log R_0^*$ values. The SRY shows higher values for $\log R_0^*$ approximately 0.36 while the most sever conditions, $2.30 < \log R_0^* < 2.60$, gave lower SRY values. The SRY decrease can be attributed to the fast hydrolysis of hemicelluloses of barley straw while this decrease is limited by resisting the hydrolysis lignin fraction and the slow hydrolysis of the crystalline cellulose fraction. Moreover, some of the produced soluble sugar degradation products (furfural and 5-hydroxymethylo-furfural) can be further degraded to insoluble byproducts (humic substances, etc.).

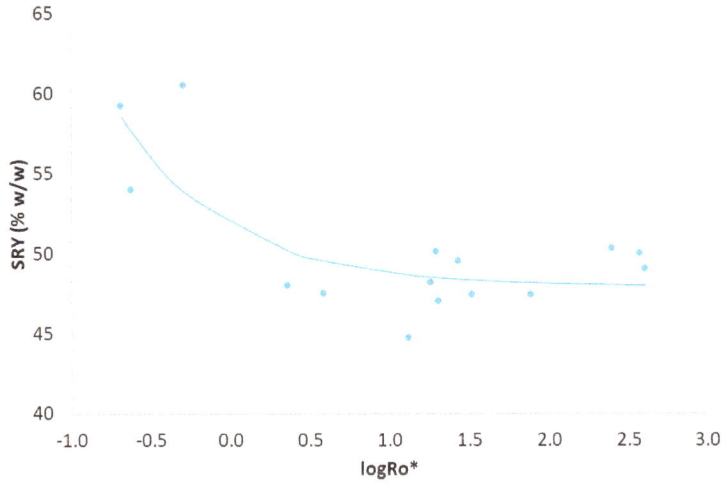

Figure 2. SRY of the ACWT-treated material as affected by the CSF.

In Table 2 the HHV and SRY values are presented, while the relationship between HHV and ACWT severity using a second-order polynomial function is shown in Figure 3. The equation with the best fighting among many others was as follows:

$$HHV = 0.5028x^2 + 0.0087x + 19.613 \tag{6}$$

where $x = \log R_0^*$. The correlation coefficient was $R = 0.7891$ and the SEE was 5.097. The increase in the HHV can be explained by the increase of the carbon content in the ACWT material, as well as by the ash percentage decreasing. In addition, it can be explained by the lignin percentage increase and the hemicelluloses percentage decreasing with regard to the pretreated barley straw.

Table 3 shows total ash content of ACWT barley straw compared to the untreated one. A decrease in ash content was observed from 8.4 for the untreated straw to 5.5% wt. for the treated one. At this point, the CSF of the ACWT was $\log R_0^* = 2.87$. Öhman et al. [46] also observed the decrease in ash content of hydrolysis residue. In addition, Jenkins et al. [47] also recognized that water-leaking biomass has decreased ash concentration. Acid hydrolysis disrupts the structure of lignocellulose. As a result, minerals presented in the biomass were released into the soluble processing fluid. Therefore, the decrease in ash content can be attributed to the disordered cell structure and the water treatment. A reduction in the total ash content can be accomplished by applying acid hydrolysis to dissolve the minerals during the process.

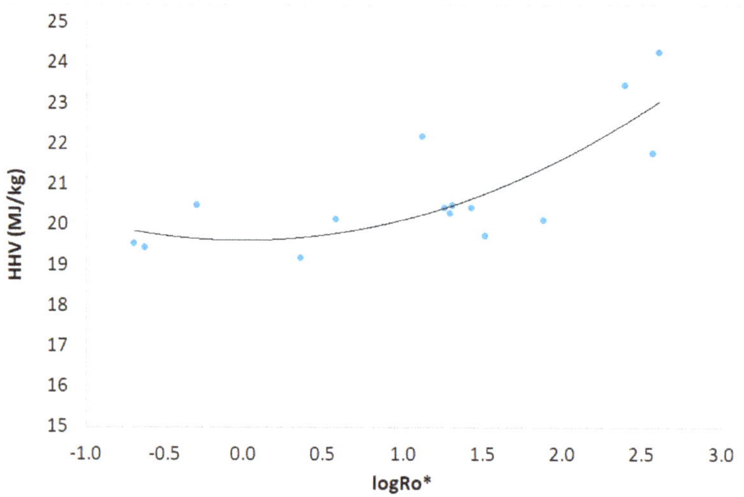

Figure 3. HHV vs. CSF in logarithmic form.

Table 3. Proximate and ultimate analysis of the original and ACWT-treated (at optimized conditions based on HHV) barley straw.

		Original Barley Straw (% wt. Dry Basis)	ACWT Barley Straw at Optimized Conditions Based on HHV (% wt. Dry Basis)	Methods
Proximate analysis	Moisture	6.00	5.10	ISO 18134-1
	Volatile Matter	74.30	72.30	ISO 18123
	Ash	8.40	5.50	ISO 18122
	Fixed Carbon	17.3	22.2	by difference
Ultimate analysis	C	45.53	52.51	ISO 16948
	H	5.50	5.79	ISO 16948
	N	0.99	0.85	ISO 16948
	S	0.12	0.14	ISO 16994
	O	47.86	40.71	by difference

In the present work, sulfur content increased after pretreatment, from 0.11% for untreated barley straw to 0.14 for the pretreated one (see Table 3). The difference is not significant, so acid hydrolysis can be applied for high sulfur containing lignocellulosics like barley straw [47].

The enhancement factor (EF) is defined by the following equation:

$$EF = HHV_t/HHV_o \quad (7)$$

where HHV_t represents the HHV for ACWT barley straw and HHV_o the HHV for original straw. The energy yield (EY) can be calculated as follows:

$$EY = EF \cdot SRY \quad (8)$$

In Figure 4, the EF and the EY of ACWT barley straw vs. the CSF in logarithmic form are presented. The curves representing the calculated values of EF and EY (i.e., the theoretical values) occurred from the application of Equations (5) and (6), respectively, using the same parameters as in Figure 3. Moreover, in Figure 4a we can notice that the EF increases when the CSF increases. On the other hand, in Figure 4b, EY demonstrates improved values either for low CSF values or for higher CSF values, but not for moderate conditions. It must be observed that taking into account both Figures 3 and 4, we get high HHV, EF and EY only for high CSF values, i.e., severe conditions. The increase in EF is a result of the increase in HHV, as can be seen from Equation (7). The initial decreasing of the EY can be explained by the significant decreasing of SRY and the insignificant increasing of EF for relatively low CSF values. On the other hand, the consequent increasing of EY can be explained by the significant increase of EF compared to the negligible SRY decreasing for higher CSF values.

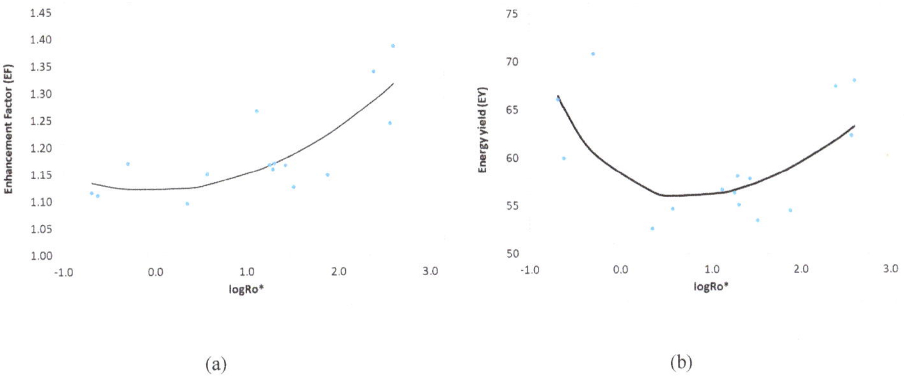

(a) (b)

Figure 4. ACWT barley straw (**a**) enhancement factor (EF) and (**b**) energy yield (EY) vs. CSF in logarithmic form.

3.2. *Ultimate and Proximate Analysis Results*

Ultimate and proximate analysis results of the barley straw compositions are presented in Table 3. The purpose of the analysis was mainly to determine whether carbon appeared during the ACWT process. In the untreated barley straw, carbon and oxygen were 45.53% and 47.86%, respectively. After ACWT pretreatment (200 °C, 0.0035 N SA concentration, 25 min), the carbon content increased to 52.51% (optimal conditions) and the oxygen decreased to 40.71%. Zanzi et al. [48], Angles et al. [40] and Iroba et al. [49] obtained similar results from the ultimate analysis of the WT-treated lignocellulosic biomass feedstock. The C content of their steam-exploded samples increases while lignin condenses and carbonizes. The lignin condensation was with loss of H_2O and reduction of O content. H and O losses were also reported due to the formation of H_2O, CO and CO_2. The observed increased N content was dropped by the increasing temperature and time of WT. The change in C, H and O content was significant. Reactions occurring during hydrolytic treatment generally produce volatiles and gases with a low energy density until they evaporate, increasing the energy density of the residue material by making it rich. The results obtained from the present ultimate and proximate analysis are consistent with the above literature results, demonstrating that carbonization occurs at higher temperatures and times in combination with degradation of holocellulose and destruction of the lignocellulosic matrix.

The increase in the HHV of the pretreated straw relative to the increase in its values of hydrogen, carbon and sulfur, as well as the reduction of oxygen, are in agreement with previous studies [44]. In the present work, the HHV of the ACWT-treated barley straw samples were calculated using the C, H and N results presented in Table 3. The following Friedl et al. [50] equation was used for these calculations:

$$HHV = 3.55C^2 - 232C - 2230H + 51.2C{\cdot}H + 131N + 20600 \quad (9)$$

The HHV for the untreated and ACWT-treated barley straw, estimated by Equation (9), was 18.1 and 21.0 MJ/kg, respectively. The experimentally measured HHV was 17.5 of untreated barley straw and 24.3 MJ/kg for ACWT-treated material. Consequently, the estimated HHV was sufficiently close to the experimental measurements. Moreover, the estimated EF was 1.16, also not so close to the experimental one that was 1.39.

3.3. Simulation Based on RSM

In this work we performed 15 experimental trials as the Box–Behnken scheme demands. The results were fitted by an RSM model using a second-order polynomial equation. The model variables, where *A* temperature, *B* time and *C* SA concentration. The suggested polynomial was fitted to the ACWT experimental data. The equation parameters were estimated by multiple linear regressions analysis. The confidence level was 99%. The optimal conditions for ACWT of barley straw were found. The polynomial equation used herein was as follows:

$$\text{HHV} = 20.397 + 1.18A + 0.585B + 0.4138C + 0.2775AB - 0.38AC - 0.0575BC + 1.0967A^2 - 1.1008B^2 + 0.7367C^2 \quad (10)$$

By solving this model using a partial differential equation (PDE), three 3D graphs were developed.

In Table 4 the effect of the three independent *A*, *B* and *C* variables on the response variables *P*, *T* and *F* is presented by ANOVA of the RSM. The coefficient of determination (R^2) equaled 0.9035, showing sufficient fitting of Equation (9) to the experimental data. Joglekar and May [51] demands $R^2 > 0.80$ for the reliable model based on its parameters. The R^2 of the model presented herein has shown that it adequately represents the true relationship between the predicted and observed values of the variables under consideration. When the R^2 value is 0.9035 it indicates that 90.3% of the volatility is explained by the model and only 9.70% is due to luck. Model regression analysis showed that the effect of the temperature variable was significant, while the time and SA concentration variable had no effect on ACWT.

Table 4. RSM coefficients and P, T and F-values for the polynomial model.

Factor	Name	Coefficients	*P*	T	F
Const	Constant	20.40	0.0000	45.76	-
A	Temperature	1.184	0.0075	4.336	18.81
B	Time	0.5850	0.0850 *	2.143	4.593
C	SA concentration	0.4137	0.1900 *	1.515	2.230
AB	-	0.2775	0.5044 *	0.7189	0.5168
AC	-	−0.3800	0.3701 *	−0.9844	0.9691
BC	-	−0.0575	0.8874 *	−0.149	0.0222
AA	-	1.097	0.0413	2.729	7.450
BB	-	−1.101	0.0408	−2.739	7.507
CC	-	0.7366	0.1262 *	1.833	3.361

* not significant.

ANOVA of the performance index, including the probability value *P*-value, the value on the F distribution F-statistic or F-value, as well as the value of the T-distribution, T-statistic or T-value, shows that the high value of F indicates that the proposed model is valid and low *P* values indicate that the model is significant. Given the F value for the temperature, which was higher compared to the other two, it implies that the increase in the reaction temperature has very strong effect on the experiment. The appropriate T value (4.3368) gives the signal to noise ratio and when the value is greater than 4 is generally desirable [52]. When *P* becomes less than 0.05, it also indicates the significance of the model terms. On the other hand, values above 0.10 means that these terms are not significant. In addition, the RSM provided a sum of squares (SS) of 27.90, degrees of freedom (DF) for the regression of 9 and 5 for the residual; a mean square (MS) of 3.100 for the regression and 0.5960 for the residual, and a significance of F (SigF) of 0.0421. ANOVA analysis produced values within the range of experimental values, confirming that the RSM model can simulate the HHV of ACWT-treated barley straw.

In Figure 5, three 3D-graphs illustrate the RSM results as a function of two of the independent variables, while the third variable remains constant. The conditions 180 °C, 20 min and 22.5 mM SA concentration were chosen as constant values as their combination gives a moderate HHV value.

The relationships between the time and temperature, temperature and SA concentration and the time and SA concentration are presented in Figure 5a–c, respectively. There was no significant effect of the combination of variables individually, namely the reaction time and the concentration of sulfate on the response value.

3.4. Optimization

The optimal conditions for ACWT of barley straw were determined by using Quantum XL commercial software. The findings were comparable to those of CSF methodology. The highest value of the HHV experimentally achieved at the expected by the Quantum XL software set point and it is achieved when relatively severe pretreatment conditions (200 °C, 20 min and 35 mol/m^3 SA concentration) were used (Table 5). Under these optimal conditions, $\log R_0^* = 2.87$ and SRY = 38% were estimated. The experimental result agreed with the optimal values predicted by the RSM model giving HHV = 25.5 MJ/kg. Both simulation approaches (CSF methodology and RSM) provided useful information on CSF and exact experimental parameters to optimize the HHV of ACWT-treated material.

Table 5. Optimal conditions (set point) for maximized HHV predicted by the RSM polynomial model.

Name	Factor	Range		Set Point
		Low	High	
Temperature	A	160	200	200
Time	B	0	40	25
SA concentration	C	10	35	35

3.5. SEM Results

The surface of the solid residue of untreated barley straw (Figure 6a,c,e) and the ACWT-treated one Figure 6b,d,f) was observed by a SEM, in the order to observe the physical changes in the structure of barley straw. The porous structure of the surface of solid residue is created during the ACWT process due to the production of volatile substances. As the process continues, the pores and cracks are appearing on the surface of the solid residue. The porous structure and the cracks of solid residue can be clearly shown on the SEM images. When the temperature increases at 200 °C, the hydrolysis process enters the second phase and most of organic materials are released gradually, resulting from a great loss of mass and the formation of the major hydrolysis products.

As the temperature rises, the volatile matter gradually evaporates, creating pores and transforming the surface of solid residue to be concave. The surface of straw is found to have more irregular porous structures in the higher hydrolysis temperature, and changes in surface morphology is observed (Figure 6b,d,f). This evidence indicates that the structure of the torrefied biofuel is downgraded or lit due to high temperature.

This results in the formation of an irregular structure of solid residue, destroying its uniformity. The solid residue, created from high temperature, is more fragile compared to the untreated straw and it cannot withstand the pressure due to its fragile structure.

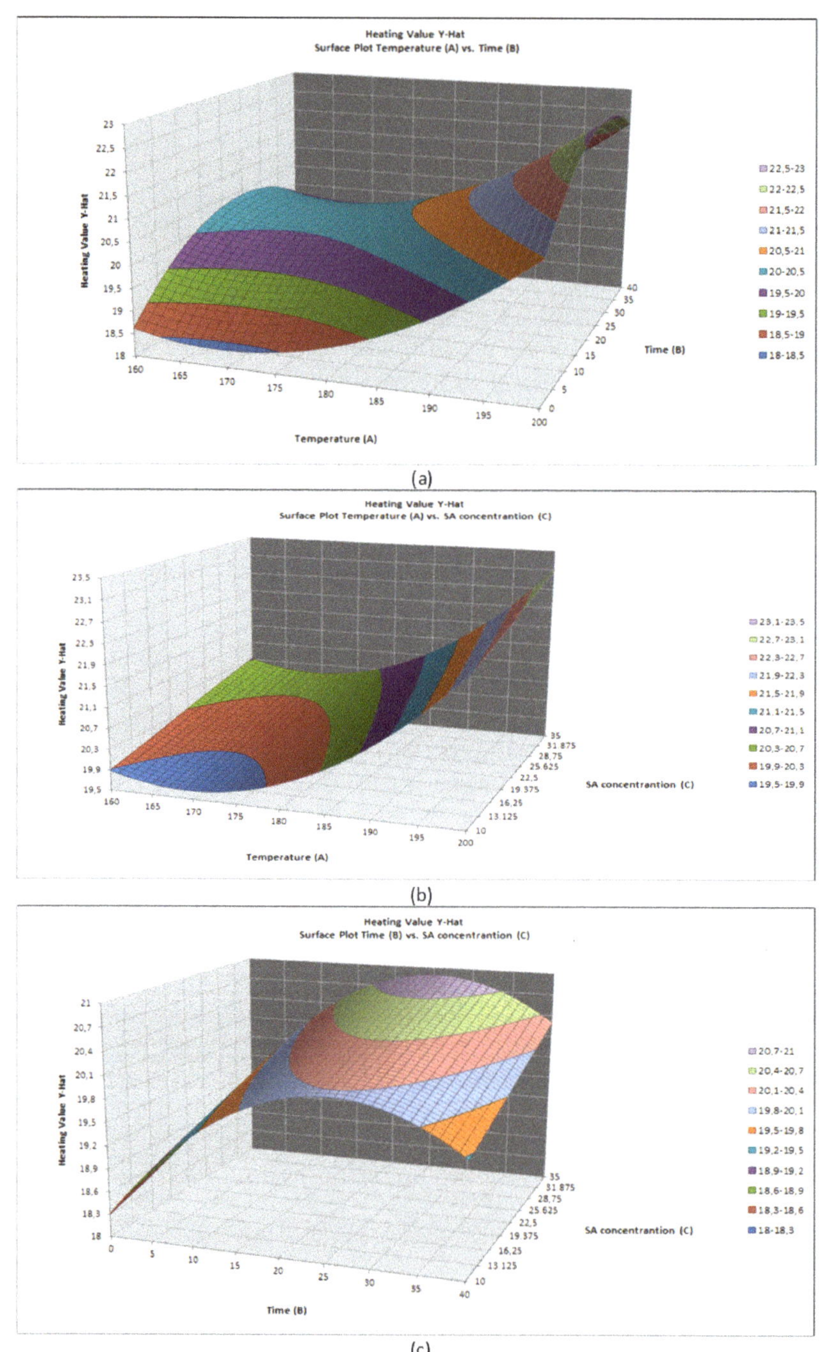

Figure 5. HHV vs. temperature–time (**a**), temperature–SA concentration (**b**) and SA concentration–time (**c**).

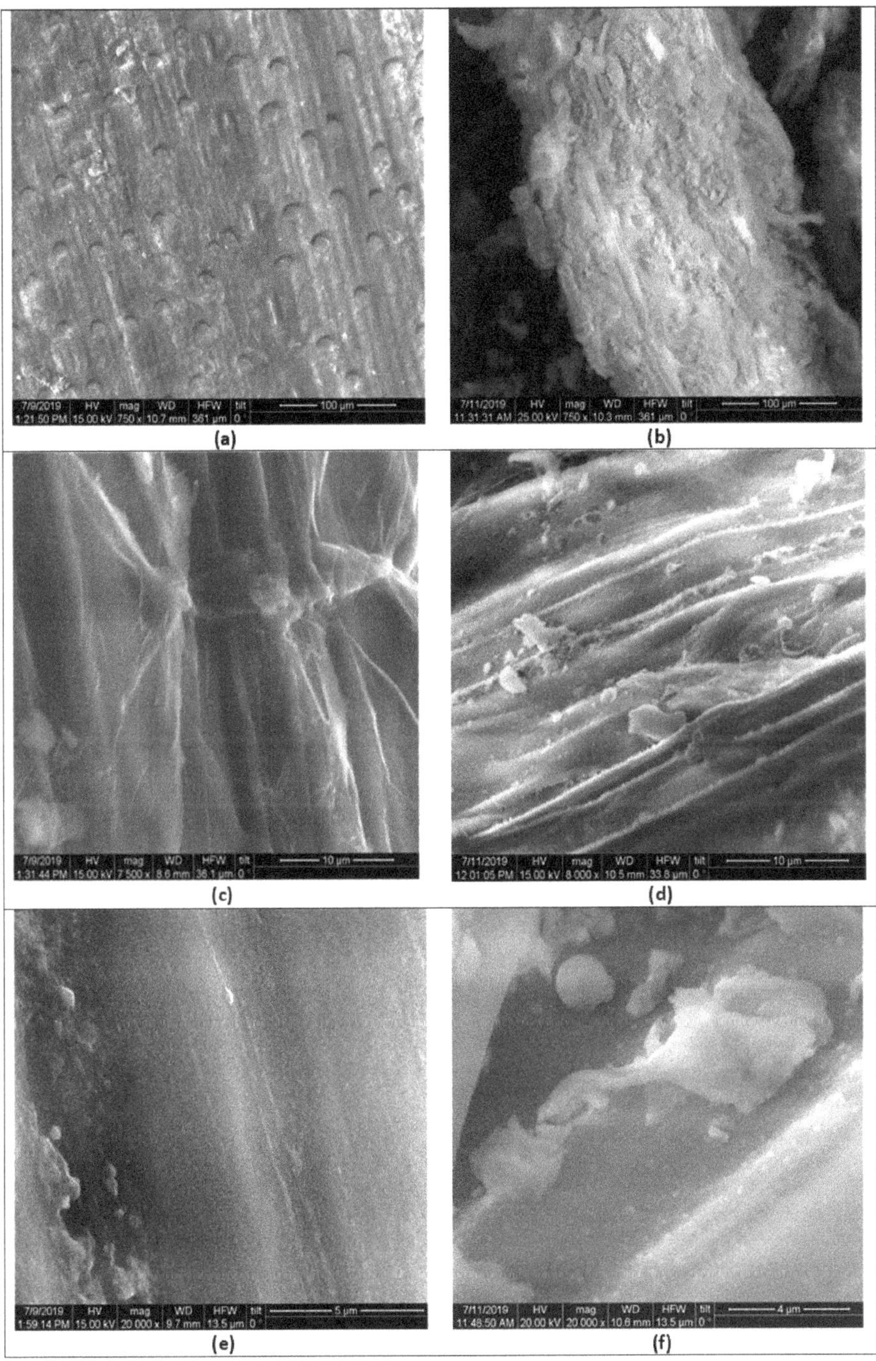

Figure 6. SEM results of original barley straw at (**a**) ×750, (**c**) ×7500 and (**e**) ×20,000 magnification, and ACWT-treated material at (**b**) ×750, (**d**) ×7500 and (**f**) ×20,000 magnification.

4. Conclusions

In this study, it was shown that ACWT has the potential to convert biomass such as barley straw into biofuel with enhanced thermal, chemical and physical fuel properties compared to raw barley straw. The results were obtained by ACWT treatment of barley straw, in a batch reactor, showing significant improvements, such as an increased HHV and lower ash content. In addition, the ACWT is appropriate for cleaner fuels production from lignocellulosic residues. Within the biorefinery concept, during the fermentable sugars production via acid hydrolysis, ACWT barley straw with enhanced HHV can be obtained as a solid byproduct. Two simulation methodologies were successfully applied for ACWT process: (i) the combined severity factor approach and (ii) the response surface methodology. The first methodology gives the CSF $R_0{}^*$ or $R_0{}^*$ (which can be achieved by various combinations of time, temperature and acid concentration) for optimized HHV, EF and EY, while the second methodology determines exactly the optimal experimental conditions (time, temperature and acid concentration). The pretreatment conditions were optimized for modified barley straw production with a maximum HHV equal to 24.3 MJ/kg, EF = 1.39 and EY = 68% wt. dry basis. The optimal ACWT temperature, time and SA concentration were 200 °C, 25 min (with regard to the isothermal period only) and a 35 mM SA concentration of the solution. The above presented simulation method, which is the combination of two methodological approaches, can be possibly used in modelling/simulation and optimization of various processes of biomass thermal, chemical and thermochemical treatment.

Author Contributions: Investigation, A.N.; methodology, P.G.; software, E.S.; supervision, D.S. All authors have read and agreed to the published version of the manuscript.

Funding: This research received no external funding.

Acknowledgments: This work has been partly supported by the University of Piraeus Research Center.

Conflicts of Interest: The authors declare no conflict of interest.

References

1. Johnsson, F.; Kjärstad, K.; Rootzén, J. The threat to climate change mitigation posed by the abundance of fossil fuels. *Clim. Policy* **2019**, *19*, 258–274. [CrossRef]
2. Kumar, R.; Strezov, V.; Weldekidan, H.; He, J.; Singh, S.; Kan, T.; Dastjerdi, B. Lignocellulose biomass pyrolysis for bio-oil production: A review of biomass pre-treatment methods for production of drop-in fuels. *Renew. Sustain. Energy Rev.* **2020**, *123*, 109763. [CrossRef]
3. Möllersten, K.; Yan, J.; Moreira, J.R. Potential market niches for biomass energy with CO_2 capture and storage—Opportunities for energy supply with negative CO_2 emissions. *Biomass Bioenergy* **2003**, *25*, 273–285. [CrossRef]
4. He, C.; Tang, C.; Li, C.; Yuan, J.; Tran, K.Q.; Bach, Q.V.; Qiua, R.; Yang, Y. Wet torrefaction of biomass for high quality solid fuel production: A review. *Renew. Sustain. Energy Rev.* **2018**, *91*, 259–271. [CrossRef]
5. Hrnčič, M.K.; Kravanja, G.; Knez, Ž. Hydrothermal treatment of biomass for energy and chemicals. *Energy* **2016**, *116*, 1312–1322. [CrossRef]
6. Tekin, K.; Karagöz, S.; Bektaş, S. A review of hydrothermal biomass processing. *Renew. Sustain. Energy Rev.* **2014**, *40*, 673–687. [CrossRef]
7. Bach, Q.-V.; Skreiberg, Ø. Upgrading biomass fuels via wet torrefaction: A review and comparison with dry torrefaction. *Renew. Sustain. Energy Rev.* **2016**, *54*, 665–677. [CrossRef]
8. Fan, S.; Xu, L.H.; Namkung, H.; Xu, G.; Kim, H.T. Influence of Wet Torrefaction Pretreatment on Gasification of Larch Wood and Corn Stalk. *Energy Fuels* **2017**, *31*, 13647–13654. [CrossRef]
9. Kostas, E.T.; Beneroso, D.; Robinson, J.P. The application of microwave heating in bioenergy: A review on the microwave pre-treatment and upgrading technologies for biomass. *Renew. Sustain. Energy Rev.* **2017**, *77*, 12–27. [CrossRef]
10. Demey, H.; Melkior, T.; Chatroux, A.; Attar, K.; Thiery, S.; Miller, H.; Grateau, M.; Sastre, A.M.; Marchand, M. Evaluation of torrefied poplar-biomass as a low-cost sorbent for lead and terbium removal from aqueous solutions and energy co-generation. *Chem. Eng. J.* **2019**, *361*, 839–852. [CrossRef]

11. Acharya, B.; Sule, I.; Dutta, A. A review on advances of torrefaction technologies for biomass processing. *Biomass Convers. Biorefin.* **2012**, *2*, 349–369. [CrossRef]
12. Chen, W.; Peng, J.; Bi, X.T. A state-of-the-art review of biomass torrefaction, densification and applications. *Renew. Sustain. Energy Rev.* **2015**, *44*, 847–866. [CrossRef]
13. Artiukhina, E.; Grammelis, P. Torrefaction of Biomass Pellets: Modeling of the Process in a Fixed Bed Reactor. *Int. J. Chem. Mol. Eng.* **2015**, *9*, 12.
14. Artiukhina, E.; Grammelis, P. Modeling of biofuel pellets torrefaction in a realistic geometry. *Therm. Sci.* **2016**, *20*, 156. [CrossRef]
15. Isemin, R.; Mikhalev, A.; Klimov, D.; Grammelis, P.; Margaritis, N.; Kourkoumpas, D.S.; Zaichenko, V. Torrefaction and combustion of pellets made of a mixture of coal sludge and straw. *Fuel* **2017**, *210*, 859–865. [CrossRef]
16. Binder, J.B.; Raines, R.T. Fermentable sugars by chemical hydrolysis of biomass. *Proc. Natl. Acad. Sci. USA* **2010**, *107*, 4516–4521. [CrossRef]
17. Taherzadeh, M.J.; Karimi, K. Acid-based hydrolysis processes for ethanol from lignocellulosic materials: A review. *Bioresources* **2007**, *2*, 472–499.
18. Román, S.; Libra, J.; Berge, N.; Sabio, E.; Ro, K.; Li, L.; Ledesma, B.; Alvarez, A.; Bae, S. Hydrothermal carbonization: Modeling, final properties design and applications: A review. *Energies* **2018**, *11*, 216. [CrossRef]
19. Román, S.; Ledesma, B.; Álvarez, A.; Coronella, C.; Qaramaleki, S.V. Suitability of hydrothermal carbonization to convert water hyacinth to added-value products. *Renew. Energy* **2020**, *146*, 1649–1658. [CrossRef]
20. Chen, W.H.; Kuo, P.C. Torrefaction and co-torrefaction characterization of hemicellulose, cellulose and lignin as well as torrefaction of some basic constituents in biomass. *Energy* **2011**, *36*, 803–811. [CrossRef]
21. Runge, T.; Wipperfurth, P.; Zhang, C. Improving biomass combustion quality using a liquid hot water treatment. *Biofuels* **2013**, *4*, 73–83. [CrossRef]
22. Kruse, A.; Funke, A.; Titirici, M.M. Hydrothermal conversion of biomass to fuels and energetic materials. *Curr. Opin. Chem. Biol.* **2013**, *17*, 515–521. [CrossRef] [PubMed]
23. Rodríguez, A.; Moral, A.; Sánchez, R.; Requejo, A.; Jiménez, L. Influence of variables in the hydrothermal treatment of rice straw on the composition of the resulting fractions. *Bioresour. Technol.* **2009**, *100*, 4863–4866. [CrossRef]
24. Zhang, C.; Ho, S.-H.; Chen, W.-H.; Xie, Y.; Liu, Z.; Chang, J.-S. Torrefaction performance and energy usage of biomass wastes and their correlations with torrefaction severity index. *Appl. Energy* **2018**, *220*, 598–604. [CrossRef]
25. Nhuchhen, D.R.; Basu, P.; Acharya, B. A Comprehensive Review on Biomass Torrefaction. *Int. J. Renew. Energy Biofuels* **2014**, *2014*, 1–56. [CrossRef]
26. Wang, X.; Wu, J.; Chen, Y.; Pattiya, A.; Yang, H.; Chen, H. Comparative study of wet and dry torrefaction of corn stalk and the effect on biomass pyrolysis polygeneration. *Bioresour. Technol.* **2018**, *258*, 88–97. [CrossRef]
27. Gan, Y.Y.; Ong, H.C.; Chen, W.H.; Sheen, H.K.; Chang, J.S.; Chong, C.T.; Ling, T.C. Microwave-assisted wet torrefaction of microalgae under various acids for coproduction of biochar and sugar. *J. Clean. Prod.* **2020**, *253*, 119944. [CrossRef]
28. Sidiras, D.K.; Nazos, A.G.; Giakoumakis, G.E.; Politi, D.V. Simulating the Effect of Torrefaction on the Heating Value of Barley Straw. *Energies* **2020**, *13*, 736. [CrossRef]
29. Brasch, D.J.; Free, K.W. Prehydrolysis-kraft pulping of Pinus radiata grown in New Zealand. *Tappi* **1965**, *48*, 245–248.
30. Overend, R.; Chornet, E. Fractionation of lignocellulosics by steam-aqueous pretreatments. *Philos. Trans. R. Soc. Lond. B Biol. Sci.* **1987**, *321*, 523–536.
31. Chum, H.H.; Johnson, D.K.; Black, S.K.; Overend, R.P. Pretreatment-catalyst effects and the combined severity parameter. *Appl. Biochem. Biotechnol.* **1990**, *13*, 24–25. [CrossRef]
32. Abatzoglou, N.; Chornet, E.; Belkacemi, K.; Overend, R.P. Phenomenological kinetics of complex systems: The development of a generalized severity parameter and its application to lignocellulosics fractionation. *Chem. Eng. Sci.* **1992**, *47*, 1109–1122. [CrossRef]
33. Lloyd, T.A.; Wyman, C.E. Combined sugar yields for dilute sulfuric acid pretreatment of corn stover followed by enzymatic hydrolysis of the remaining solids. *Bioresour. Technol.* **2005**, *96*, 1967–1977. [CrossRef] [PubMed]

34. Kabel, M.A.; Bos, G.; Zeevalking, J.; Voragen, A.G.J.; Schols, H.A. Effect of pretreatment severity on xylan solubility and enzymatic breakdown of the remaining cellulose from wheat straw. *Bioresour. Technol.* **2007**, *98*, 2034–2042. [CrossRef] [PubMed]
35. Sidiras, D.; Batzias, F.; Ranjan, R.; Tsapatsis, M. Simulation and optimization of batch autohydrolysis of wheat straw to monosaccharides and oligosaccharides. *Bioresour. Technol.* **2011**, *102*, 10486–10492. [CrossRef] [PubMed]
36. Weinwurm, F.; Turk, T.; Denner, J.; Whitmore, K.; Friedl, A. Combined liquid hot water and ethanol organosolv treatment of wheat straw for extraction and reaction modeling. *J. Clean. Prod.* **2017**, *165*, 1473–1484. [CrossRef]
37. Ferreira, S.L.C.; Bruns, R.E.; Ferreira, H.S.; Matos, G.D.; David, J.M.; Brandão, G.C.; da Silva, E.G.P.; Portugal, L.A.; dos Reis, P.S.; Souza, A.S.; et al. Box-Behnken design: An alternative for the optimization of analytical methods. *Anal. Chim. Acta* **2007**, *597*, 179–186. [CrossRef]
38. Bezerra, M.A.; Santelli, R.E.; Oliveira, E.P.; Villar, L.S.; Escaleira, L.A. Response surface methodology (RSM) as a tool for optimization in analytical chemistry. *Talanta* **2008**, *76*, 965–977. [CrossRef]
39. Box, G.E.P.; Wilson, K.B. On the experimental attainment of optimum conditions. *J. R. Stat. Soc. Ser. B* **1951**, *13*, 1–45. [CrossRef]
40. Salapa, I.; Katsimpouras, C.; Topakas, E.; Sidiras, D. Organosolv pretreatment of wheat straw for efficient ethanol production using various solvents. *Biomass Bioenergy* **2017**, *100*, 10–16. [CrossRef]
41. Angles, M.N.; Reguant, J.; Garcia-Valls, R.; Salvado, J. Characteristics of lignin obtained from steam-exploded softwood with soda/anthraquinone pulping. *Wood Sci. Technol.* **2003**, *37*, 309–320. [CrossRef]
42. Toussaint, B.; Excoffier, G.; Vignon, M.R. Effect of steam explosion treatment on the physico-chemical characteristics and enzymic hydrolysis of poplar cell wall components. *Anim. Feed Sci. Technol.* **1991**, *32*, 235–342. [CrossRef]
43. Heitz, M.; Capek-Menard, E.; Koeberle, P.G.; Gagne, J.; Chornet, E.; Overend, R.P.; Taylor, J.D.; Yu, E. Fractionation of *Populus tremuloides* at the pilot plant scale: Optimization of steam pretreatment conditions using the STAKE II technology. *Bioresour. Technol.* **1991**, *35*, 23–32. [CrossRef]
44. Demirbas, A. Biomass Resource Facilities and Biomass Conversion Processing for Fuels and Chemicals. *Energy Convers. Manag.* **2001**, *42*, 1357–1378. [CrossRef]
45. Semhaoui, I.; Maugard, T.; Zarguili, I.; Rezzoug, S.A.; Zhao, J.M.Q.; Toyir, J.; Nawdali, M.; Maache-Rezzoug, Z. Eco-friendly process combining acid-catalyst and thermomechanical pretreatment for improving enzymatic hydrolysis of hemp hurds. *Bioresour. Technol.* **2018**, *257*, 192–200. [CrossRef] [PubMed]
46. Öhman, M.; Boman, C.; Hedman, H.; Eklund, R. Residential combustion performance of pelletized hydrolysis residue from lignocellulosic ethanol production. *Energy Fuels* **2006**, *20*, 1298–1304. [CrossRef]
47. Jenkins, B.M.; Bakker, R.R.; Wei, J. On the properties of washed straw. *Biomass Bioenergy* **1996**, *10*, 177–200. [CrossRef]
48. Zanzi, R.; Ferro, D.T.; Torres, A.; Soler, P.B.; Bjornbom, E. Biomass Torrefaction. In Proceedings of the 2nd World Biomass Conference on Biomass for Energy Industry and Climate Protection, Rome, Italy, 10–14 May 2004; In World Biomass Conference-CD-ROM Edition; WIP: Munich, Germany; Florence, Italy, 2004; pp. 859–862.
49. Iroba, K.L.; Tabil, L.G.; Sokhansanj, S.; Dumonceaux, T. Pretreatment and fractionation of barley straw using steam explosion at low severity factor. *Biomass Bioenergy* **2014**, *66*, 286–300. [CrossRef]
50. Friedl, A.E.; Padouvas, H.R.; Varmuza, K. Prediction of heating values of biomass fuel from elemental composition. *Anal. Chim. Acta* **2005**, *544*, 191–198. [CrossRef]
51. Joglekar, A.; May, A. Product excellence through design of experiments. *Cereal Foods World* **1987**, *32*, 211–230.
52. Cao, G.; Ren, N.; Wang, A.; Lee, D.J.; Guo, W.; Liu, B.; Feng, Y.; Zhao, Q. Acid hydrolysis of corn stover for biohydrogen production using Thermo-anaerobacterium thermo-saccharolyticum W16. *Int. J. Hydrogen Energy* **2009**, *34*, 7182–7188. [CrossRef]

Article

Resource Assessment and Numerical Modeling of CBM Extraction in the Upper Silesian Coal Basin, Poland

Jarosław Chećko [1], Tomasz Urych [1], Małgorzata Magdziarczyk [2] and Adam Smoliński [1,*]

1 Central Mining Institute, Plac Gwarków 1, 40-166 Katowice, Poland; jchecko@gig.eu (J.C.); turych@gig.eu (T.U.)

2 Faculty of Economics and Management, Opole University of Technology, ul. Luboszycka 7, 45-036 Opole, Poland; m.magdziarczyk@po.edu.pl

* Correspondence: smolin@gig.katowice.pl; Tel.: +48-32259-2252

Received: 27 March 2020; Accepted: 28 April 2020; Published: 1 May 2020

Abstract: The paper presents the assessment of the resources of methane considered as the main mineral in the most prospective selected areas of the Upper Silesian Coal Basin, Poland in the region of undeveloped deposits. The methane resources were estimated by means of a volumetric method at three depth levels, 1000, 1250, and 1500 m. A part of the Studzienice deposit comprising three coal seams, 333, 336, and 337, located in a methane zone was chosen for the numerical modeling of simulated methane production. The presented static 3D model has been developed using Petrel Schlumberger software. The total resources of methane in the area amount to approximately 446.5 million of Nm^3. Numerical simulations of methane production from the selected coal seams with hydraulic fracturing were conducted by means of Schlumberger ECLIPSE reservoir simulator. Based on the simulations, it was concluded that, in the first six months of the simulations, water is produced from the seams, which is connected with the decrease in the rock mass pressure. The process prompts methane desorption from the coal matrix, which in turn results in a total methane production of 76.2 million of Nm^3 within the five-year period of the simulations, which constitutes about 17% of total methane resources (GIP). The paper also presents a detailed analysis of Polish legislation concerning the activities aimed at prospecting, exploring, and extracting the deposits of hydrocarbons.

Keywords: coal seams; methane resources; numerical simulations; legal issues

---

1. Introduction

Coal bed methane (CBM) is a natural gas created as a result of the transformations of organic matter into bituminous coal, which, due to the sorption phenomena, was accumulated in different ranks of coal [1]. The sorption capacity of coal depends on various parameters, i.e., thermodynamic conditions within the coal seams, moisture, and petrographic content of coals, as well as the composition of the adsorbed gas [2–6]. The global geological resources of coal bed methane are estimated to range from 100 to 216 billion m^3 with the recoverable CBM resources estimated at 24–42 billion m^3 [7,8]. In the United States, both the recognition of exploitable resources and the production of CBM are well-developed. The recoverable resources of coal bed methane in the United States account for 9%–29% of the geological resources, whereas the exploitable ones constitute only 14% of the recoverable resources [9]. The CBM resources are subject to the same procedures concerning the prospecting, exploring, documenting, and production as the resources of conventional hydrocarbons. The feature that makes coal bed methane resources different from the conventional ones is connected with large variability of the parameters used to estimate the resources, which in consequence translates into a high level of uncertainty concerning the reported statistics. The application of deterministic methods to assess the resources is encumbered with errors; therefore, it is recommended to use probabilistic forecasting, which takes into consideration the statistical variability of parameters—for example, Monte Carlo methods [10]. Companies which carry on exploration and production activities, as well as those which produce reports on CBM resources, most often apply the petroleum resource management system (PRMS) introduced in 2007 under the auspices of international oil industry companies [11]. The PRMS classification of hydrocarbons resources allows for the assessment of the range of uncertainty concerning the predicted volume of hydrocarbons, as well as the risk assessment of commerciality. In Poland, the classification of CBM as the main mineral is based on the criteria of recoverability stipulated by the Regulation of the Minister of Environment on the geological and investment report of a hydrocarbon deposit [12].

Pre-mine drainage of methane from coal seams is an important issue concerning the safety of coal miners in operating coal mines. In recent years, there have been some works associated with the pre-mining methane extraction from coal seams using surface directional wells [13–15].

New prospects regarding the utilization of coal bed methane in Poland started to appear in the 1990s when AMOCO, TEXACO, Metanel, and Pol-Tex Methane initiated in the period of 1991–1998 intensified works on CBM extraction in the Upper Silesian Coal Basin within the framework of their concessions [16]. The extraction method consisted in simultaneous opening out of multiple gas-bearing coal seams by means of vertical wells with perforation and the use of hydraulic fracturing technology. The method proved to be unprofitable in the case of the Upper Silesian Coal Basin due to the small capacity of methane production during the tests, which did not guarantee the project cost effectiveness. The most important geological factor responsible for the poor performance was the low permeability of the coal seams in the Upper Silesian Coal Basin [16].

Previous research was carried out by means of vertical wells, whereas currently there is a trend of applying the technology of directional (horizontal) drilling for the opening out of gas-bearing coal seams. Particularly good results were obtained for directional drilling combined with under-balanced drilling [17]. The horizontal drilling ensures large drainage area, whereas under-balanced drilling allows preventing the damage of the near-well zone [18]. Thanks to the application of the above drilling methods in the United States, Australia, and Canada, it has been possible to produce gas from poorly permeable coal seams whose exploitation, similarly to the CBM production in the Upper Silesian Coal basin, was previously considered uneconomical [19]. The first evaluations of CBM resources in the Upper Silesian Coal Basin were made in the early 1990s [20]. The underlying reason to provide such evaluation was the process of granting concessions for the CBM resources in Poland. The in-situ resources of methane were estimated by means of the volumetric method, which is commonly applied in the United States [21].

In consecutive years, various institutions (Table 1) recalculated the estimates of CBM resources. All the estimates encompassed in situ resources and were performed using the volumetric method; in the subsequent estimates, only the boundary conditions were changed.

Table 1. Coal bed methane (CBM) resources in the Upper Silesian Coal Basin.

Institution	CBM Resources (Million Nm^3)
Polish Geological Institute [20]	365
Katowice Geological Company [22]	no less than 300–320
Central Mining Institute [23]	442

Methane resources were estimated for the basic zone of methane-bearing coal seams according to the current criteria of recoverability, i.e., up to the depth of 1500 m for coal seams of a minimal thickness of 0.6 m and an average methane content higher than 2.5 m^3/Mg_{daf}. Yet another verification of the resources of coal bed methane, considered as the main mineral, took place in 2006. In this report, the resources of CBM were estimated at 124.5 billion m^3, including the areas of operating coal mines below the depth of 1000 m, the closed seams, and undeveloped areas (Table 2).

Table 2. Estimated recoverable CBM resources within the boundaries of the selected area.

Type of Seam	CBM Resources (Million Nm^3)
Seams in operating coal mines	42,050.2
Seams in closed coal mines	7184.2
Undeveloped seams	61,958.0
Separated areas	13,347.7
Total	**124,540.2**

In Table 2, methane resources were estimated according to the volumetric method (GIP)-based on the verification of methane resources in active coal mines (coal mine methane—CMM), undeveloped coal deposits (coal bed methane—CBM), and abandoned coal mines (AMM). Due to the fact that, in Poland, the geological criteria for the estimation of CMM are different from those for CBM, the methane resources in Table 2 take into account different boundary conditions for resource estimation: (a) for CBM—the depth of 1500 m, seams with thickness >0.6 m, methane capacity >4.5 m^3/Mg_{daf}; and (b) for CMM—the depth of documented coal seams in the mine area, thickness >0.2 m, and methane capacity of 2.5 m^3/Mg_{daf}.

Three-dimensional geological modeling of hydrocarbon resources plays a prominent role in the processes of prospecting, exploring, developing, and exploiting hydrocarbon accumulations. In this respect, Schlumberger Petrel software constitutes a tool for developing geological-structural models that enable the integration of geological and geophysical data, as well as the information concerning reservoir engineering. The package of petrel reservoir engineering core modules, combined with an ECLIPSE industry-reference simulator, provides a set of numerical solutions for forecasting the dynamic reservoir processes and for designing plans for the exploration, development, and production. The modeling of bituminous coal seams using Schlumberger Petrel software enables to perform numerical simulations of coal bed methane extraction by means of directional drilling from the surface [19,20], as well as computer simulations of enhanced methane recovery combined with injecting CO_2 into the coal seams-enhanced coal bed methane (ECBM) [24–31].

The aim of this article is the evaluation of the resources of coal bed methane considered as the main mineral for 19 undeveloped seams within the area of the Upper Silesian Coal Basin, as well as the selection of the most prospective areas in terms of methane resources and the estimation of in-place methane resources (GIP) at several depth intervals. Additionally, the study encompasses the development of numerical simulations of methane extraction from selected coal deposits in order to estimate the recoverable methane resources and calculate the possible value of recovery factor.

The methane resources were estimated by means of the volumetric method (GIP—Gas In Place). Three areas of the Upper Silesian Coal Basin were selected as the most prospective. The analysis allowed assessing the distribution of CBM resources in the vertical profile in the adopted intervals (a) from the roof of the zone to 1000 m, (b) 1000–1250 m, and (c) 1250–1500 m for the three prospective areas. The total recoverable reserves of CBM considered as the main mineral in the first region equal 17,243.3 million Nm^3, in the second region—12,447.4 million Nm^3, and in the third region—3887 million Nm^3.

The Studzienice deposit was selected for the numerical modeling of CBM extraction. A structural parameter model of the coal-bearing Upper Carboniferous formations was developed, and the reservoir parameters were analyzed. The static model was supplemented with detailed reservoir parameters, as well as the thermodynamic properties of reservoir gases and fluids.

Within the framework of the study, numerical simulations have been performed for the process of methane extraction from selected coal seams by means of a multilateral well combined with hydraulic fracturing to enhance the production. In addition, numerical simulations of methane extraction using directional wells drilled from the surface were carried out, and a numerical model of the selected coal deposits was developed.

2. Materials and Methods

2.1. The Estimation of Coal Bed Methane Resources

The resources of coal bed methane considered as the main mineral were estimated for 19 undeveloped seams within the area of the Upper Silesian Coal Basin. The resources were estimated for the designated boundaries of the seams areas, similar to the resources of methane as an accessory mineral within the seams of operating coal mines. The total recoverable resources of methane as the main mineral amount to 60,125 million Nm^3 [32]. For the purpose of the estimates of CBM resources, the following boundary conditions were adopted:

- the thickness of the seams >0.3 m,
- the depth—1600 m, and
- the methane content >4.5 m^3/Mg_{daf}—only in the basic zone of methane-bearing coal seams.

The methane resources were estimated by means of the volumetric method (GIP—Gas In Place) at three depth levels—1000, 1250, and 1500 m (see Figure 1).

The volumetric method is one of the simplest methods of estimating resources; it requires the relatively smallest recognition of the deposit, especially in terms of identifying reservoir parameters; however, it is generally used only to calculate geological resources and, after adopting geological criteria, also the anticipated economic and subeconomic ones. The calculation of CBM gas resources (GIP—Gas In Place) using the volumetric method is based on the following formula:

$$GIP = A \cdot H \cdot M_{avg} \cdot \gamma \quad (1)$$

where

GIP = methane resources (m^3),
A = calculation area of methane resources (m^2),
H = thickness of dry ash free coal (m),
M_{avg} = average content of hydrocarbons (m^3/Mg_{daf}), and
γ = density (Mg/m^3).

For the purposes of further research, three areas of the Upper Silesian Coal Basin were selected as the most prospective in terms of methane resources. The analysis will allow assessing the distribution of CBM resources in the vertical profile in the adopted intervals for estimating resources (a) from the roof of the zone to 1000 m, (b) 1000–1250 m, and (c) 1250–1500 m for the three prospective areas. Next, a numerical model of the selected coal deposit was developed, and numerical simulations of the methane extraction using directional wells drilled from the surface to assess the potential developed resources were carried out.

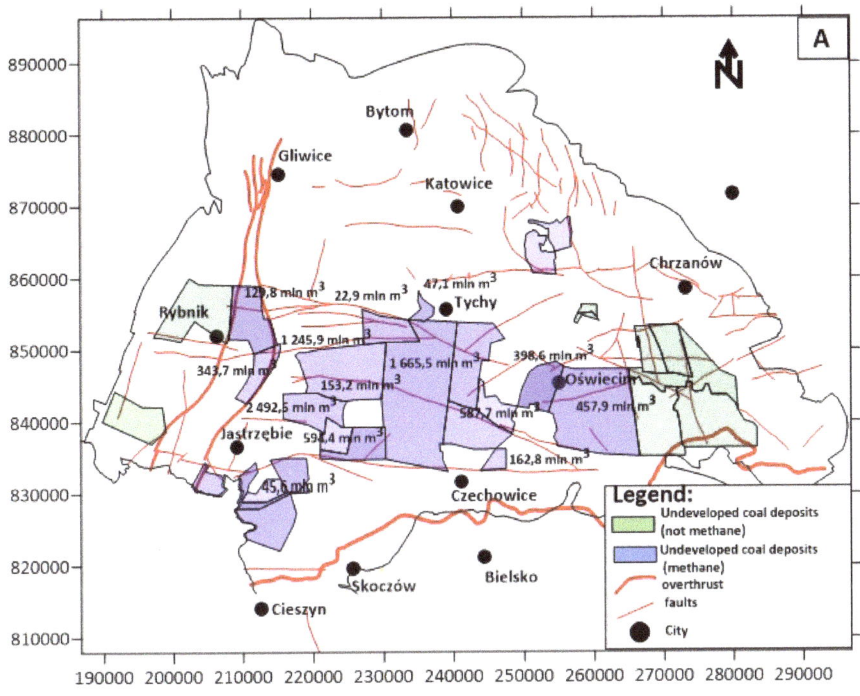

(a)

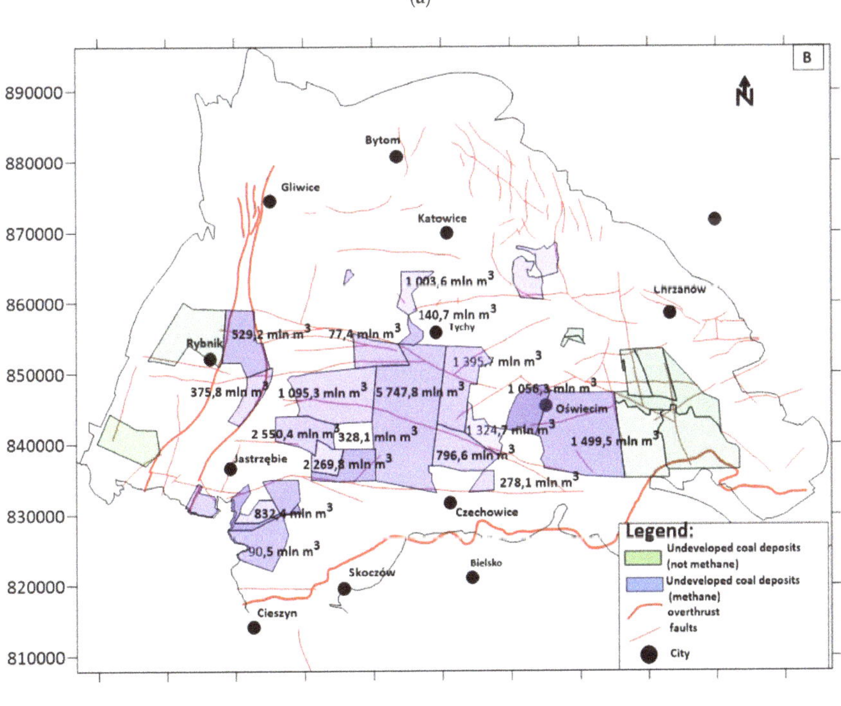

(b)

Figure 1. *Cont.*

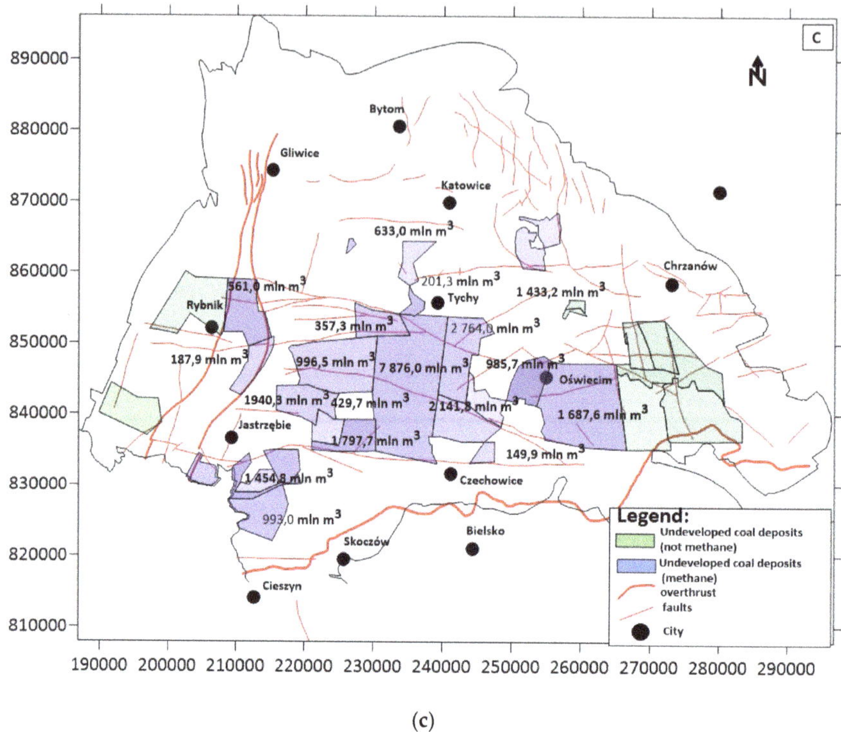

(c)

Figure 1. Maps of coal bed methane (CBM) resources occurring in undeveloped areas of the Upper Silesian Coal Basin, Poland for three depths: (**a**) up to 1000 m, (**b**) 1000–1250 m, and (**c**) 1250–1500 m.

2.2. Numerical Modeling of CBM Extraction

The ECLIPSE Simulator allows for the double porosity in the model. The model consists of two interconnected systems representing the coal matrix and the cleat system. Methane is stored by means of sorption in a poorly permeable coal matrix characterized by varied porosity in comparison to the cleat system where the phenomenon of desorbed gas flow takes place. Accordingly, in double-porosity models, the number of layers is doubled, whereas the calculations during the simulation are conducted for a double number of cells.

The model of gas adsorption on coal for the different components is described in ECLIPSE 300 by means of the extended Langmuir isotherm [33]. The adsorption capacity is a function of the pressure and free-gas phase composition. For each of the gases, it is required to introduce Langmuir isotherm parameters, i.e., the Langmuir volume constant V_i and the Langmuir pressure constant P_i. These parameters are typically determined based on experiments. Different isotherms can be used in different regions of the field. The multicomponent adsorption capacity is calculated by means of the following equation:

$$L(p, y_1, y_2, \ldots)_i = \theta \frac{P_s}{RT_s}\left(V_i \frac{y_i \frac{p}{P_i}}{1 + \sum_{j=1}^{nc} y_j \frac{p}{P_j}}\right) \quad (2)$$

where

θ = scaling factor,

P_s = pressure at standard conditions,

R = universal gas constant,

T_s = temperature at standard conditions,

V_i = Langmuir volume constant for component i,

P_i = Langmuir pressure constant for component i,

y_i = hydrocarbon mole fraction in the gas phase for component i, and

p = pressure.

For the special case of a single component, the extended Langmuir isotherm is identical to the usual Langmuir isotherm giving the storage capacity as a function of pressure only:

$$L(p) = \theta \frac{P_s}{RT_s}\left(\mathbf{V}\frac{\frac{p}{\mathbf{P}}}{1+\frac{p}{\mathbf{P}}}\right) \tag{3}$$

where $\mathbf{V}$ is the maximum storage capacity for the gas, referred to as the Langmuir volume constant, and $\mathbf{P}$ is the Langmuir pressure constant. The constants used in the extended Langmuir formulation can hence be estimated from a series of single-component gas experiments.

Time-dependent diffusion in ECLIPSE 300, i.e., the diffusive flow between the matrix and the fracture, is given by:

$$F_i = DIFFMF \cdot D_{c,\,i} \cdot S_g \cdot RF_i \cdot (m_i - \rho_c\, L_i) \tag{4}$$

where

m_i = molar density in the coal matrix, and

$DIFFMF$ = matrix fracture (or multi-porosity) diffusivity.

It is possible to scale the adsorption capacity, keyword LANGMULT, by a factor for each cell in the grid. Typically, this can be used to account for differences in ash or moisture contents.

ρ_c = rock density (coal density),

$D_{c,I}$ = diffusion coefficient (coal) component i,

RF_i = readsorption factor component i,

S_g = gas saturation for a desorption value of unity is used, and

$\rho_c\, L_i$ = equilibrium molar density of adsorbed gas.

The matrix-fracture diffusivity is given by:

$$DIFFMF = DIFFMMF \cdot VOL \cdot \sigma \tag{5}$$

where

$DIFFMMF$ is the multiplying factor input,

VOL is the coal volume, and

σ is the factor to account for the matrix-fracture interface area per unit volume.

Often the component's sorption time is a quantity that is easier to obtain than the diffusion coefficients. For desorption, we write the flow as:

$$F_i = (VOL/\tau_i)\,(m_i - \rho_c\, L_i) \tag{6}$$

where

$\tau_i = 1/(D_{c,i} \cdot DIFFMMF \cdot \sigma)$ is called the sorption time.

The parameter controls the time lag before the released gas enters the coal fracture system.

The sorption times are given by the diffusion coefficients, *DIFFCBM*, and the matrix-fracture interface area, SIGMA, together with the multiplying factor *DIFFMMF*. If the sorption times are known, a value of unity can be assigned to σ and *DIFFMMF*. The diffusion coefficients can then be assigned to the reciprocal of the sorption times [33].

Simulator ECLIPSE 300 requires predetermining the initial gas concentration in the coal by means of inputting the gas volume to the mass of the base rock (sm^3/kg). The ECLIPSE software defines the sm^3/kg unit as a cubic meter of gas (pressure of 1 atm = 1013.25 hPa and temperature of 15.56 °C) per one kilogram of coal under in-situ conditions.

The presented static 3D model has been developed using Petrel Schlumberger 2010.1 software [34]. The above model constituted the foundation for creating a dynamic model encompassing the simulations of the process of CBM extraction by means of directional drilling technology combined with hydraulic fracturing of the rock mass to enhance methane production. The numerical simulations were performed by means of the 2011.3 compositional version of the ECLIPSE simulator with the option of coal bed methane [33]. The numerical model was constructed for the selected area of research encompassing the region of an undeveloped Studzienice deposit. During the first phase of the study, a static model of the whole Studzienice play, displaying the following documented seams: 211, 213/1, 214/1, 214/1-2, 215/2, 301, 303, 308, 310, 312, 315, 318, 320/1, 321, 324, 328, 333, 336, and 337 and taking into consideration the occurrence of potential faults, was constructed. Within the model, the coal-bearing formations of productive carboniferous demonstrate considerable a tectonic engagement of the rock mass (Figure 2).

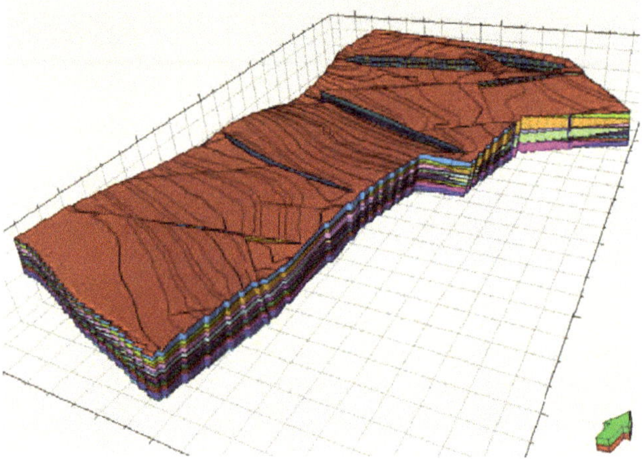

Figure 2. Structural model of bituminous coal seams.

The area of the modeled region equals 55.5 km^2. The static model was used for developing dynamic simulation models for the selected seams 333, 336, and 337 of the high-methane content zone. Within one tectonic block, the exploration and production wells were located. The first designed well, Studzienice V, was a vertical one and performed the function of an exploration well by means of which the analyzed coal seams were drilled. The next well consisted of three branched directional boreholes (Studzienice 1H, Studzienice 2H, and Studzienice 3H) drilled from a vertical section of the well, which enabled the simulation of methane exploitation processes. The depths of the deposition of the seams within the selected model block were in the ranges of 925–1023 m, 1007–1102 m, and 1021–1116 m for seams 333, 336, and 337, respectively. Figure 3 presents the deposition of seam 337 within the selected tectonic block.

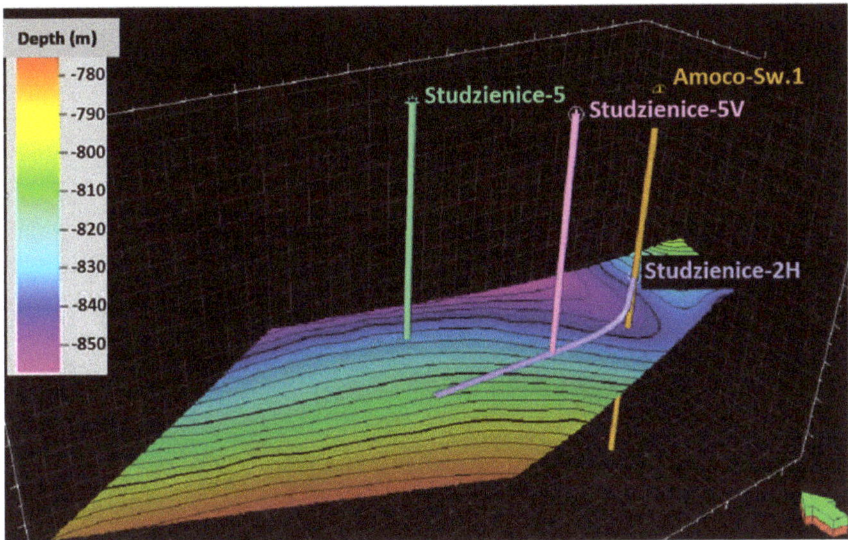

Figure 3. The deposition of seam 337 within the selected tectonic block.

The area of the modeled region equals 2.62 km^2. The thickness of the modeled region ranges from 70.5 to 113 m. A horizontal resolution of a 10 × 10-m interpolation grid was applied. Figure 4 demonstrates the developed model, which comprises 5 layers and 219,375 cells. The parameters of the numerical model are listed in Table 3.

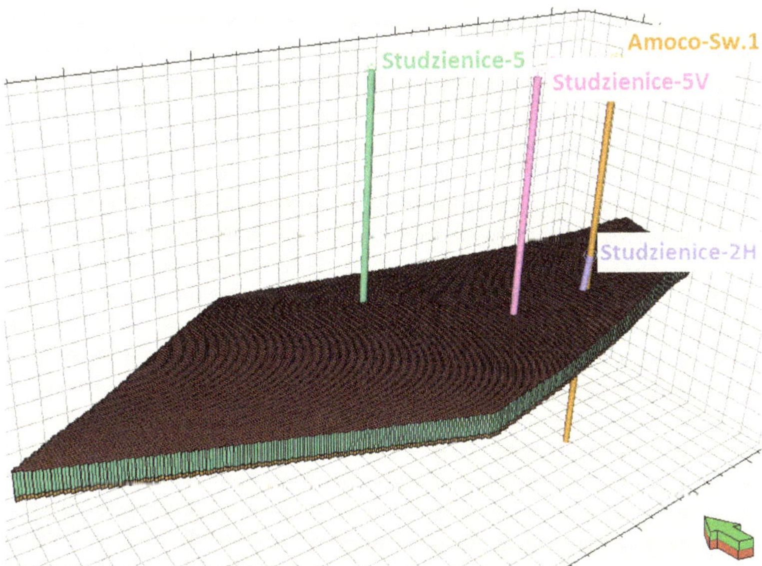

Figure 4. Simulation model of CBM extraction from seams 333, 336, and 337.

Table 3. Structural parameters of the simulation model for CBM extraction.

Calculation Interval (Depth, M)	Recoverable Reserves of CBM Considered as the Main Mineral (Million Nm^3)
Average coal seam thickness (m)	1.5
The area of the model (km^2)	2.62
Horizontal resolution of the interpolation grid (m)	10 × 10
Vertical resolution of the model (m)	1.5
Fracture porosity (%)	0.5
Permeability X, Y, and Z (mD)	1.5

In the developed numerical model, a series of numerical simulations have been performed for the process of methane extraction from selected coal seams by means of a multilateral well combined with hydraulic fracturing to enhance production. For the purpose of simulating the process of CBM extraction, a compositional version of ECLIPSE simulator with the option of coal bed methane was used. This version of the simulator takes into consideration the major mechanisms responsible for the flow of water and gas within the coal seam, namely, the desorption of gas from the coal matrix into the cleat system, diffusion in the coal matrix according to Fick law, and Darcy's flow in the fractures [33]. The ECLIPSE 300 Simulator used in the study allows for the double porosity in the model. Therefore, the model consists of two interconnected systems representing the coal matrix and the cleat system. Methane is stored by means of sorption in a poorly permeable coal matrix characterized by varied porosity in comparison to the cleat system where the phenomenon of desorbed gas flow takes place. Additionally, gas adsorption isotherms described in ECLIPSE 300 by means of an extended Langmuir isotherm are applied in the model [30]. Besides the parameters of the extended Langmuir isotherm, the developed model also incorporates the coal density, methane diffusion coefficient, and minimum production pressure. The ECLIPSE software defines the sm^3/kg unit as a cubic meter of gas (at the pressure of 1 atm = 1013.25 hPa and temperature of 15.56 °C) per one kilogram of coal under in-situ conditions [33].

In the developed model, laboratory data obtained during the execution of the Reduction of CO_2 emission by means of CO_2 storage in coal seams in the Silesian Coal Basin of Poland (RECOPOL) project and a pilot CBM project executed by TEXACO were used [16,34–37]. Detailed parameters for developing the model are compiled in Table 4.

Table 4. Parameters used for modeling the CBM extraction.

Parameter	Seams 333, 336, and 337
Coal density, kg/m^3	1330.00
CH_4 Diffusion coefficient, m^2/d	0.0000685
Initial pressure, bar	115–120
Min. Production pressure, bar	5.00
333 coal seam depth, m	925–1023
336 coal seam depth, m	1007–1102
337 coal seam depth, m	1021–1116
Extended Langmuir Isotherm Parameters	
CH_4 volume V_L, sm^3/kg	0.0205
CH_4 pressure P_L, bar	42.00
CO_2 volume V_L, sm^3/kg	0.0320
CO_2 pressure P_L, bar	19.03

3. Legal Aspects of the Activities Concerning the Prospection, Exploration, and Extraction of CBM in Light of the Act on Geological and Mining Law

Under the provisions of the Act of 9 June 2011, Geological and Mining Law (i.e., of 4 April 2019, the Journal of Laws of 2019, item 868; hereafter: Geological and Mining Law), the deposits of hydrocarbons—including, in particular, coal bed methane—constitute mining ownership regardless of the place of their occurrence. The deposits of hard coal, methane as an associated mineral, lignite, metal ores, excluding bog iron ores, native metals, radioactive ores, native sulfur, rock salt, sylvinite, potassium-magnesium salts, gypsum and anhydrite, precious stones, rare-earth elements, noble gases, as well as the deposits of curative water, thermal water, and brine are covered by mineral rights wherever they occur. Mineral rights are exercised solely by the State Treasury. Under the provisions of legal acts, the State Treasury may, to the exclusion of other persons, use the object of the mineral rights or dispose of its rights solely by establishing a mining usufruct which shall be established by way of an agreement in writing. In light of the said Act, the activity consisting in the prospection, exploration, and production of hydrocarbon deposits may be carried on exclusively after obtaining a concession. Due to the character of the State Treasury, the entitlements of the State Treasury provided by mineral rights with respect to the above activities are exercised by competent concession-granting authorities, i.e., bodies entitled by law to grant decisions on the basis of which it is possible to undertake an activity which requires a concession. According to the legislator, the prospection of hydrocarbon deposits means performing geological development works with a view to establishing and preliminarily documenting the existence of a deposit. The exploration of hydrocarbon deposits means carrying out geological development works in the area of preliminarily documented hydrocarbon deposits. The extraction of hydrocarbons from deposits means conducting the hydrocarbon extraction, as well as preparing the extracted hydrocarbons for transport and their transport within the area of a mining plant. The legislator devoted a separate chapter in the Act on Geological and Mining Law to provide regulations concerning the issue of granting concessions for the prospection, exploration, and production of hydrocarbon deposits (Act on Geological and Mining Law, Chapter III). The procedure of applying for concessions for the prospection, exploration, and production of hydrocarbon deposits, unlike in the case of other concessions, is preceded by an eligibility procedure to assess the ability of the interested entity to carry out the activities in the scope of the prospecting for and exploration of hydrocarbon deposits and the extraction of hydrocarbons from deposits. Within the framework of the eligibility procedure that is carried out by the Minister of the Environment, it is determined whether the entity intending to apply for a concession for the prospecting for and exploration of a hydrocarbon deposit and the extraction of hydrocarbons from a deposit, or a concession for the extraction of hydrocarbons from a deposit, remains under corporate control of a third-party state, entity, or citizen of a third-party state, and, if so, if such control can pose a threat to national security. In such a case, a third-party state is understood to be any state other than a Member State of the European Union, Member State of the European Free Trade Agreement (EFTA), or a Member State of the North Atlantic Treaty Organization. Corporate control is interpreted as all forms of obtaining, directly or indirectly, entitlements, which, separately or jointly, given all legal and factual circumstances, make it possible for the entity to exert decisive influence on the entity interested in conducting activities within the scope of prospecting for and exploration of a hydrocarbon deposit and the extraction of hydrocarbons from a deposit or a concession for the extraction of hydrocarbons from a deposit. An example of circumstances that enable to exert influence on the entity interested in conducting said activities is the case where the said entity holds, directly or indirectly, a majority of votes at the general meeting or is entitled to appoint or dismiss a majority of the members of the management board or supervisory board. The aforementioned eligibility procedure is conducted upon the application of the entity interested in conducting activities within the scope of prospecting for and exploration of a hydrocarbon deposit and the extraction of hydrocarbons from a deposit or a concession for the extraction of hydrocarbons from a deposit. The legislator requires from the minister responsible for the environment to keep a list of the entities to which a decision on the positive assessment in the eligibility procedure has been issued. It should be underlined that the decision on the positive assessment in the eligibility procedure constitutes a mandatory prerequisite that entitles the entity to which it has been issued to submit a tender in the tender procedure for obtaining a concession to prospect for and explore hydrocarbon deposits and to extract hydrocarbons from deposits or a concession to extract hydrocarbons from deposits. A concession to prospect for and explore hydrocarbon deposits and to extract hydrocarbons from deposits or a concession to extract hydrocarbons from deposits may be granted by way of a tender procedure but also upon the request of the interested entity. The tender procedure should have an objective and nondiscriminatory character, and it should give priority to the best systems of prospecting for and the exploration of hydrocarbon deposits or the extraction of hydrocarbons from deposits.

The legislator cautioned that the tender procedure should be based on the following criteria:

(1) experience in the activity concerning the prospecting for and the exploration of hydrocarbon deposits or the extraction of hydrocarbons from deposits in order to ensure the safety of the conducted activities in terms of the protection of human and animal life and health, as well as the protection of the environment;
(2) technical capabilities to carry out the activity in the scope of prospecting for and the exploration of hydrocarbon deposits and the extraction of hydrocarbons from deposits or the extraction of hydrocarbons from deposits, respectively, and, in particular, having relevant technical, organizational, logistics, and staff resources, including the cooperation in the development and implementation of innovations in prospecting for, exploring, and extracting hydrocarbons from deposits with scientific institutions dealing with the exploration of the geological structure of Poland, as well as analytics, technology, and methodology of prospecting for deposits that take into consideration the specificity of Polish geological conditions and apply to these conditions;
(3) financial capabilities providing due guarantee of performing the activity of prospecting for and the exploration of hydrocarbon deposits and the extraction of hydrocarbons from deposits or the extraction of hydrocarbons from deposits, respectively, and, in particular, the sources and manners of financing the intended activities, including the share of own funds and funds originating from third-party capital;
(4) proposed technology of performing geological development works, including geological works or mining works;
(5) scope and schedule of the proposed geological development works, including geological works or mining works; and
(6) scope and schedule of the obligatory sampling resulting from geological works, including drilling cores.

The amount of the fee for the establishment of the mining usufruct, due at the stage of prospecting and exploring, may constitute an additional criterion when two or more tenders submitted by the entities interested in prospecting for and the exploration of hydrocarbon deposits or the extraction of hydrocarbons from deposits obtain the same score.

The concession for prospecting for and the exploration of a hydrocarbon deposit and the extraction of hydrocarbons from a deposit or the concession for the extraction of hydrocarbons from a deposit should define at least the following issues:

(1) the type and the manner of performing the indented activity;
(2) the area within which the intended activity is to be pursued;
(3) the period of the concession validity;
(4) the date of the commencement of the activities covered by the concession and the specific conditions—in particular, those pertaining to public safety and health, environmental protection, or rational management of the deposit;
(5) the boundaries of the mining district and mining area; and
(6) the minimum degree of the utilization of the deposit resources and the measures necessary for the purpose of the rational management of the deposit.

The concession to prospect for, to explore hydrocarbon deposits, and to extract hydrocarbons from deposits is granted for a fixed period; however, not shorter than 10 years and not longer than 30 years. Under the provisions of the concession, the entrepreneur is granted an exclusive right to carry out the activities covered with the concession in the area defined therein. In addition, the concession granting authority concludes a mining usufruct agreement with the entity.

It is worth noting that one of the obligations imposed by the legislator on the entrepreneur who has been granted the concession for prospecting for, exploring hydrocarbon deposits, and extracting hydrocarbons from deposits is to provide the state geological service with current parameters of the extraction of hydrocarbons from the deposit and to notify the concession granting authority of such communication.

4. Results and Discussion

4.1. The Estimation of Coal Bed Methane Resources

The first of the selected prospective areas is located between the fault zone of Żory-Suszec-Jawiszowice in the North and the fault zone of Bzie-Czechowice in the South of Poland. Within the boundaries of the designated area, there are the undeveloped deposits Warszowice-Pawłowice Północ and Pawłowice-rej, the sub-marginal deposits Studzionki-Mizerów and the Southern section of the Kobiór-Pszczyna deposit. The region is considered to be relatively well-recognized in terms of methane conditions. The fact that, within the said region, there occur CBM resources in the zone of secondary methane accumulation estimated at 5.18 billion Nm^3 that constitutes an additional argument in favor of regarding the region as the most promising one [38]. Sixty-five wells are located in the area; the wells were used to make tests concerning the methane content of the seams by means of a one-phase vacuum degassing in the so-called ball containers, a method designed by Katowice Geological Company (*Katowickie Przedsiębiorstwo Geologiczne*). Another eight wells located in the same region were used to perform the tests of free desorption using the US Bureau of Mine (USBM) method developed in the United States, which enables to measure the content of residual methane and to estimate the volume of desorbable methane. The roof of the basic zone of methane bearing seams (the roof of the series) of methane content >4.5 m^3/Mg_{daf} is deposited at the depth of approximately 750 m in the East up to the depth of 1200 m in the region of the Pawłowice deposit. The estimated recoverable reserves in the first area (Figure 5) are presented in Table 5.

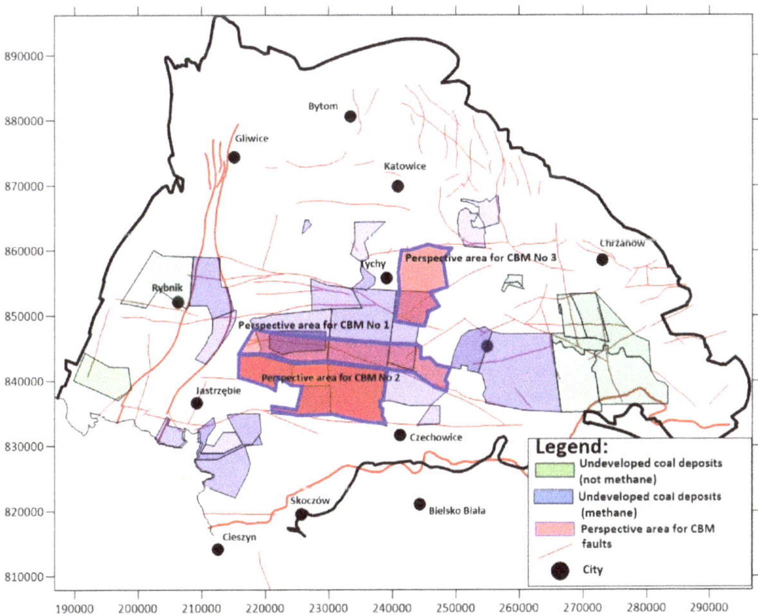

Figure 5. Prospective regions in terms of methane production using directional drilling from the surface set against the undeveloped areas of the Upper Silesian Coal Basin.

Table 5. Recoverable reserves of CBM considered as the main mineral in the first prospective region.

Calculation Interval (Depth, M)	Recoverable Reserves of CBM Considered as the Main Mineral (Million Nm^3)
From the roof of the zone to 1000 m	1050.3
1000–1250 m	8361.6
1250–1500 m	7831.4
Total	**17,243.3**

In Tables 5–7, methane resources were estimated according to the volumetric method (GIP, Gas-In-Place) in the adopted intervals for estimating resources (a) from the roof of the zone to 1000 m, (b) 1000–1250 m, and (c) 1250–1500 m for three perspective areas taking into account the following boundary conditions: depth 1500 m, thickness of coal seams >0.6 m, and methane capacity >4.5 m^3/Mg_{daf}.

Table 6. Recoverable reserves of CBM considered as the main mineral in the second prospective region.

Calculation Interval (Depth, M)	Recoverable Reserves of CBM Considered as the Main Mineral (Million Nm^3)
From the roof of the zone to 1000 m	2674.4
1000–1250 m	4034.8
1250–1500 m	5738.1
Total	**12,447.4**

Table 7. Recoverable reserves of CBM considered as the main mineral in the third prospective region.

Calculation Interval (Depth, M)	Recoverable Reserves of CBM Considered as the Main Mineral (Million Nm^3)
From the roof of the zone to 1000 m	28.4
1000–1250 m	1123.7
1250–1500 m	2734.9
Total	**3887.0**

The second prospective area is located in the hanging wing of the fault zone Żory-Suszec-Jawiszowice (Figure 5). The area encompasses the Międzyrzecze deposit and the Southern section of the Studzienice deposit, as well as the Central section of the Kobiór-Pszczyna deposit. The region is considered to be most prospective in the Eastern part (Międzyrzecze deposit), which is indicated by a considerable dynamic of methane desorption. There are 35 wells located in the area to perform tests concerning the methane content of the seams by means of the above-described method developed by the Katowice Geological Company and another four wells to perform the tests using the USBM method. The roof of the zone lies at depths of the order of 750–1000 m. The estimated recoverable reserves in the second prospective area are compiled in Table 6.

The third selected prospective area encompasses the documented CBM deposit Lędziny and the Northern section of the Studzienice deposit (Figure 5). There are 15 wells located in the area to perform tests concerning the methane content of the seams by means of the method developed by the Katowice Geological Company and another three wells to perform the tests using the USBM method. The roof of the CBM zone lies at the depth interval ranging from 950 to over 1100 m.

The estimated recoverable reserves in the third prospective area are shown in Table 7.

4.2. Numerical Simulation of CBM Extraction

In all of the simulation scenarios examined, hydraulic fracturing combined with directional drilling and casing string perforation was applied. The purpose of the casing string perforation was to achieve the maximum production rate of the well. The particular scenarios differed in terms of the hydraulic fracturing. For all of the production wells, hydraulic fracturing simulations were carried out for a number of depth intervals within the period of six to eight months from the commencement of the injection.

During the first six months, it was clearly observed for each of the simulations that, predominantly, water was produced from the seams. In this period, it was found that the rock mass pressure decreases, which is caused by water production from the seam. The process prompts methane desorption from the coal matrix. The simulation covered the period of five years; the daily production rates of methane and water are presented in Figure 6. The simulated methane extractions from seams 333, 336, and 337 are facilitated by means of horizontal wells in particular seams—Studzienice 1H, Studzienice 2H, and Studzienice 3H, respectively. The Studzienice 5V well performs the function of monitoring the depth of the coal seam deposition and, thus, constitutes a benchmark for the horizontal wells. Due to the low value of coal seams' permeability of the analyzed deposit, a multilateral production well was used in the model.

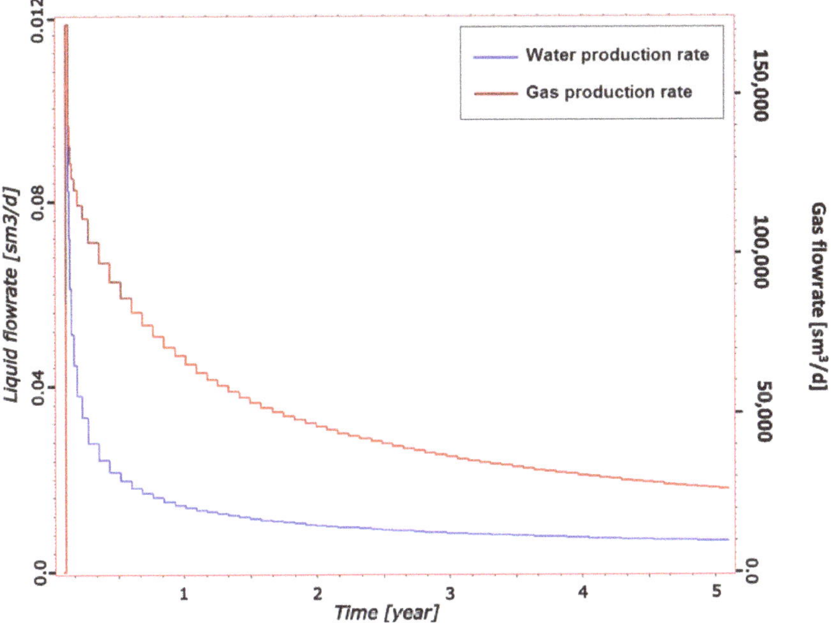

Figure 6. Daily production rates of water and methane during the simulated extraction of CBM for the entire model of the deposit using three directional wells.

Multilateral directional drilling is a technology developed and applied for the first time in the United States in order to improve the production flow and increase the efficiency of production rates of single wells. The system of branched multilateral wells drilled into the coal seam facilitates the extraction of approximately 85%–90% of gas contained in a 5-km^2 play over a period of about 30 months [39]. So far, the systems of branched multilateral wells have been successfully applied in the development of CBM technology in Alberta Basin, Canada and San Juan Basin, the United States, as well as in Bowen/Sydney Basin, Australia [40–43]. Branched multilateral wells constitute an effective solution to achieve high production rates of the CBM extraction process. The application of the system of branched multilateral wells is beneficial for the CBM technology due to the larger extent of drainage zone. Moreover, such a system enables to achieve the maximum rate of methane production in a relatively short time, along with the decrease in water drainage from the coal seam [44].

Figure 7 presents the distribution of multilateral wells in the simulation model and the changes in the distribution of coal seam pressure during the simulation of methane extraction. The considerable decrease in the rock mass pressure was caused by the extraction of water and methane from the coal seams.

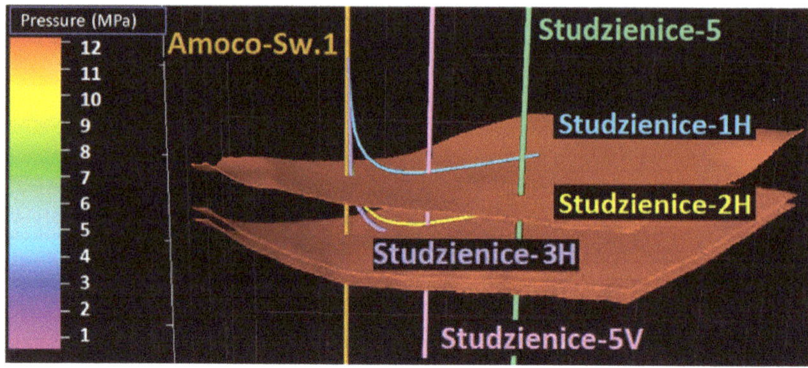

(**a**)

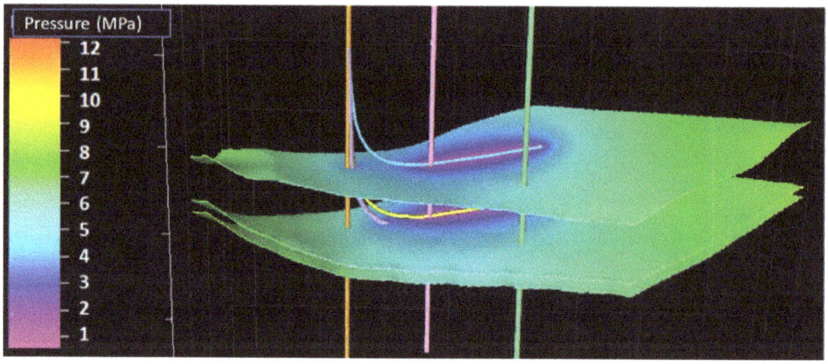

(**b**)

Figure 7. Distribution of pressure in coal seams (**a**) before the simulation and (**b**) after 5 years of methane extraction (end of the simulation).

For the assumed parameters of the deposit, the aggregate methane production rate from coal seams 333, 336, and 337 totaled approximately 76.2 million Nm^3. The rates of methane production for the particular horizontal wells are presented in Table 8, whereas Figure 8 shows the comparison of the aggregate methane production rate in the wells during the simulation.

Table 8. Rates of methane production for the particular wells.

Production Well	Methane Production Rate (Nm^3)
Studzienice-1H	23,633,812
Studzienice-2H	27,043,441
Studzienice-3H	25,513,001
Total	**76,190,254**

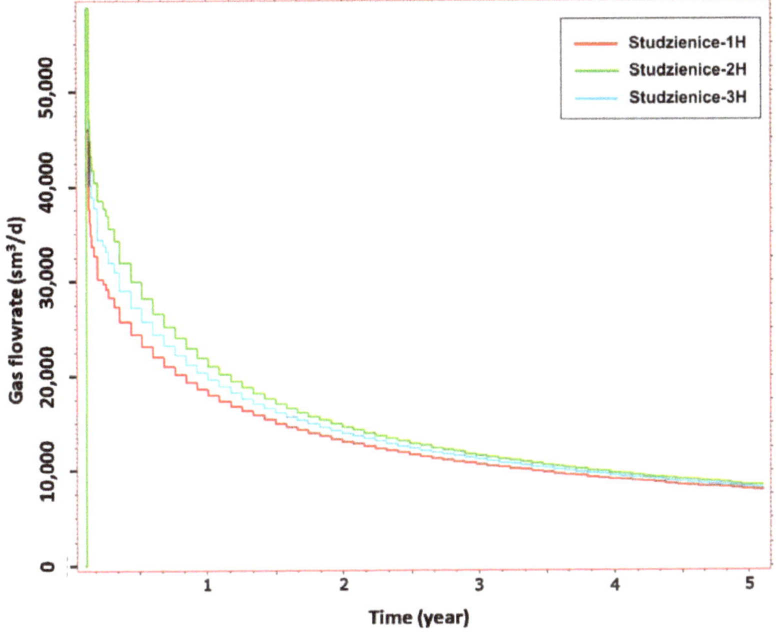

Figure 8. The comparison of the changes in methane production rates in the particular wells.

The results of the numerical simulations demonstrate that the aggregate volume of methane extracted from three coal seams totals 76.2 million Nm^3 in the five-year period of the simulation. Due to the low permeability of coal in the deposit (the assumed coal permeability in the numerical model equals 1.5 mD), the technology of multilateral wells was decided to be incorporated into the simulation model. The aggravate methane production rate for each of the branches of the multilateral well—in particular, coal seams—was in the range of 23.6–27.0 million Nm^3, whereas the average daily methane production rates ranged from 12,900 to 14,800 thousand Nm^3/d. The obtained results of the numerical simulations can be compared with the results of experimental works carried out in the Upper Silesian Coal Basin in the period of 2016–2018 within the framework of a research project Geo-Metan [45,46]. Geo-Metan was a pilot project executed in the area of an undeveloped coal deposit Międzyrzecze whose main objective was the development and refinement of Polish technologies of CBM exploration and pre-mining extraction. Within the test phase, which covered the period of 145 days, the directional well drilled into a coal seam of 0.2–0.8-mD permeability achieved a maximum flow of methane of 10,000 Nm^3/d after a hydraulic fracking procedure was applied. With the completion of water drainage, the daily production rate of methane stabilized at the level of approximately 5000 Nm^3/d. The lower values of the daily production rates achieved in the experimental works, as compared to the results of numerical simulations, may be caused by different reservoir parameters of the analyzed deposits, especially coal permeability.

5. Conclusions

The paper presents the assessment of the resources of methane considered as the main mineral in the most prospective selected areas of the Upper Silesian Coal Basin, Poland in the region of undeveloped deposits. The methane resources were estimated by means of the volumetric method (GIP) at three depth levels—1000, 1250, and 1500 m.

Three areas of the Upper Silesian Coal Basin were selected as the most prospective in terms of methane resources. The analysis allowed assessing the distribution of CBM resources in the vertical profile in the adopted intervals for estimating resources: (a) from the roof of the zone to 1000 m, (b) 1000–1250 m, and (c) 1250–1500 m for three prospective areas.

The total recoverable reserves of CBM considered as the main mineral in the first prospective region equal 17,243.3 million Nm^3, in the second region—12,447.4 million Nm^3, and in the third region—3887 million Nm^3. The Studzienice deposit was selected for the numerical modeling of CBM extraction. A structural parameter model of the coal-bearing Upper Carboniferous formations was developed, and the reservoir parameters were analyzed. The static model was supplemented with detailed reservoir parameters, as well as the thermodynamic properties of reservoir gases and fluids. Next, numerical simulations of methane extraction from the selected seams were carried out.

The above analyses were possible to perform by means of ECLIPSE software with the CBM option. The software compatibility with the petrel reservoir engineering package enabled to simulate the process of CBM extraction. The results concerning the changes in the average rock mass pressure during the simulated methane extraction were presented. The findings concerning the efficiency of a multilateral well system extracting methane from three selected coal seams were discussed.

The total resources of methane in the area amounted to approximately 446.5 million Nm^3. Based on the simulations, it was concluded that, in the first six months, water is produced from the seams, which is connected with the decrease in the rock mass pressure. The process prompts methane desorption from the coal matrix, which, in turn, results in a total methane production of 76.2 million Nm^3 within the 5five-year period of the simulations, which constitutes about 17% of total methane resources (GIP).

The results of numerical simulations confirm that the application of multilateral well systems combined with hydraulic fracturing considerably improves the efficiency of CBM extraction from seams characterized by low coal permeability. In comparison with the conventional wells, the multilateral well systems provide the possibility of optimizing the economic and technical performance of CBM extraction. In addition, the paper also presents an in-depth analysis of Polish legislation concerning the activities aimed at prospecting, exploring, and extracting the deposits of hydrocarbons in light of the binding regulations stipulated by the Act on Geological and Mining Law.

Author Contributions: Conceptualization, A.S., J.C., M.M., and T.U.; methodology, J.C., T.U., and M.M., formal analysis, A.S., T.U., M.M., and J.C.; investigation, T.U., J.C., M.M., and A.S.; resources, T.U, J.C., and A.S.; writing—original draft preparation, J.C., T.U., and M.M.; writing—review and editing, T.U., J.C., and M.M.; visualization, T.U.; and supervision, A.S. All authors have read and agreed to the published version of the manuscript.

Funding: This research was funded by Ministry of Science and Higher Education, Poland, grant number 11112018-122.

Conflicts of Interest: The authors declare no conflicts of interest.

References

1. Levine, J.R. Generation, storage and migration of natural gas in coal bed reservoirs. *Alta. Res. Counc. Inf. Ser.* **1990**, *109*, 84–130.
2. Seidle, J. *Fundamentals of Coalbed Methane Reservoir Engineering*; PennWell Books: Tulsa, OK, USA, 2011.
3. Chattaraj, S.; Mohanty, D.; Kumar, T.; Halder, G. Thermodynamics, kinetics and modeling of sorption behaviour of coalbed methane—A review. *J. Unconv. Oil Gas Resour.* **2016**, *16*, 14–33. [CrossRef]
4. Wojtacha-Rychter, K.; Smolinski, A. A study of dynamic adsorption of propylene and ethylene emitted from the process of coal self-heating. *Sci. Rep.* **2019**, *9*, 18277. [CrossRef] [PubMed]
5. Wojtacha-Rychter, K.; Howaniec, N.; Smoliński, A. The effect of coal grain size on the sorption of hydrocarbons from gas mixtures. *Int. J. Energy Res.* **2019**, *43*, 3496–3506. [CrossRef]
6. Wojtacha-Rychter, K.; Smoliński, A. Selective adsorption of ethane, ethylene, propane, and propylene in gas mixtures on different coal samples and implications for fire hazard assessments. *Int. J. Coal Geol.* **2019**, *202*, 38–45. [CrossRef]
7. Kuuskraa, V.A.; Stevens, S.H. *Worldwide Gas Shales and Unconventional Gas: A Status Report*; American Clean Skies Foundation (ACSF): Washington, DC, USA; Research Partnership to Secure Energy for America (RPSEA): Houston, TX, USA, 2009.

8. Thomas, L. *Coal Geology*, 2nd ed.; Wiley-Blackwell: Oxford, UK, 2013.
9. Hadro, J.; Spińczyk, A.; Chećko, J. *Ocena Potencjału Zasobowego MPW dla Pokładów Węgla Zalegających w strefie Wtórnego Nagromadzenia Metanu Sorbowanego Południowej Części Niecki Głównej*; CAG Państwowy Instytut Geologiczny: Warszawa, Poland, 2004.
10. Reeves, S. Unconventional gas resources to reserves—A predictive approach. In Proceedings of the AAPG Rocky Mountain Geology & Energy Resources Conference, Denver, Colorado, 9–11 July 2008.
11. Petroleum Resource Management System (PRMS). Society of Petroleum Engineers/American Association of Petroleum Geologists/World Petroleum Council/Society of Petroleum Evaluation Engineers. 2007. Available online: www.spe.org (accessed on 1 April 2020).
12. Regulation of the Minister of Environment, 2015; item 968; The Regulation of 10 July 2015; Journal of Laws of 2015. Available online: http://prawo.sejm.gov.pl/isap.nsf/download.xsp/WDU20150000968/O/D20150968.pdf (accessed on 1 April 2020).
13. Jureczka, J. Pre-mining methane extraction and capture from coal seams by using surface boreholes in the "Mysłowice Wesoła" hard coal mine project—Progress and initial assessment of works. *Bezpieczeństwo Pracy i Ochr. Środowiska w Górnictwie* **2014**, *12*, 21–26.
14. Wojtacha-Rychter, K.; Smoliński, A. Multi-component gas mixture transport through porous structure of coal. *Fuel* **2018**, *233*, 37–44. [CrossRef]
15. Wojtacha-Rychter, K.; Smoliński, A. The interactions between coal and multi-component gas mixtures in the process of coal self-heating at different various temperatures ranges: An experimental study. *Fuel* **2018**, *213*, 150–157. [CrossRef]
16. McCants, C.Y.; Spafford, S.; Stevens, S.H. Five-spot production pilot on tight spacing: Rapid evaluation of a coalbed methane block in the Upper Silesian Coal Basin, Poland. In Proceedings of the 2001 International Coalbed Methane Symposium, Tuscaloosa, AL, USA, 14–18 May 2001; pp. 193–204.
17. Shen, R.; Qiao, L.; Fu, L.; Yang, H.; Shi, W. Research and Application of Horizontal Drilling for CBM. *Soc. Pet. Eng.* **2012**. [CrossRef]
18. He, Z.; Jun-cheng, L.; Zhi-jun, Z.; Wei, L. Research on Safety Monitoring System of Bottom Hole Pressure in Coalbed Methane Underbalanced Drilling Process. *Open Fuels Energy Sci. J.* **2015**, *8*, 197–201. [CrossRef]
19. Palmer, I. Coalbed methane completions: A world view. *Int. J. Coal Geol.* **2010**, *82*, 184–195. [CrossRef]
20. Kotas, A.; Kwarciński, J.; Jureczka, J. *Calculation of Methane Resources in Coal Seams of the Upper Silesian Coal Basin*; Narodowe Archiwum Geologiczne PIG-PIB: Warszawa, Poland, 1990.
21. Ayers, W.B., Jr. Coalbed gas systems, resources, and production and a review of contracting cases from the San Juan and Powder River basins. *Am. Assoc. Pet. Geol. Bull.* **2002**, *86*, 1853–1890.
22. Pękała, Z. Occurrence of methane in the Upper Silesian Coal Basin against the background of the general geological structure of the area. *Biul. Inf. Geol.* **1992**, *1*, 8.
23. Kaziuk, H.; Bromek, T.; Chećko, J.; Chudzicka, B. Conditions of coal seams methane in the Upper Silesian Coal Basin and its resources. In Proceedings of the International Conference on the Use of Coal Bed Methane, Katowice, Poland, 5–7 October 1994.
24. Zhang, M.; Yang, Y.; Xia, Z.; Cui, Z.; Ren, B.; Zhang, W.; Lau, H.C. Best Practices in Static Modelling of a Coalbed Methane Field: An Example from the Bowen Basin in Australia. *Soc. Pet. Eng.* **2014**. [CrossRef]
25. Zhou, F.; Yao, G.; Tyson, S. Impact of geological modeling processes on spatial coal bed methane resource estimation. *Int. J. Coal Geol.* **2015**, *146*, 14–27. [CrossRef]
26. Sinayuc, C.; Shi, J.Q.; Imrie, C.E.; Syed, S.A.; Korre, A.; Durucan, S. Implementation of horizontal well CBM/ECBM technology and the assessment of effective CO_2 storage capacity in a Scottish coalfield. *Energy Procedia* **2011**, *4*, 2150–2156. [CrossRef]
27. Galas, C.; Savenko, V.; Kieke, D.E. *Simulation Study of Methane and Carbon Dioxide Migration and Leakage during Normal and Enhanced Field Operations to Recover Coal Bed Methane from Coal Seams*; Carbon Dioxide Capture for Storage in Deep Geologic Formations; Eide, L.I., Ed.; CPL Press: Totnes, UK, 2009; Volume 3.
28. Anggara, F.; Sasaki, K.; Amijaya, H.; Sugai, Y.; Setijadji, L.D. CO_2 injection in coal seams, an option for geological CO_2 storage and enhanced coal bed methane recovery (ECBM). In Proceedings of the Indonesian Petroleum Association, Thirty-Fourth Annual Convention & Exhibition, Jakarta, Indonesian, 18–20 May 2010.
29. Jureczka, J.; Chećko, J.; Krieger, W.; Kwarciński, J.; Urych, T. Prospects for geological storage of CO2 with enhanced coal bed methane recovery in the Upper Silesian Coal Basin. *Biul. Państwowego Inst. Geol.* **2012**, *448*, 117–131.

30. Chećko, J.; Urych, T.; Magdziarczyk, M.; Smoliński, A. Research on the Processes of Injecting CO_2 into Coal Seams with CH_4 Recovery Using Horizontal Wells. *Energies* **2020**, *13*, 416. [CrossRef]
31. Koteras, A.; Checko, J.; Urych, T.; Magdziarczyk, M.; Smolinski, A. An Assessment of the Formations and Structures Suitable for Safe CO2 Geological Storage in the Upper Silesia Coal Basin in Poland in the Context of the Regulation Relating to the CCS. *Energies* **2020**, *13*, 195. [CrossRef]
32. Kwarciński, J. *Weryfikacja Bazy Zasobowej Metanu Pokładów Węgla Jako Kopaliny Głównej Na Obszarze Górnośląskiego Zagłębia Węglowego*; Narod. Arch. Geol. PIG-PIB: Warszawa, Poland, 2006.
33. *ECLIPSE Reservoir Simulation Software*; Simulation Software Manuals 2011.1; Schlumberger: Houston, TX, USA, 2011.
34. *Petrel Seismic-to-Simulation Software*; Version 2010.1; Schlumberger: New York, NY, USA, 2010.
35. Arri, L.E.; Yee, D.; Morgan, W.D.; Jeansonne, N.W. Modeling Coalbed Methane Production with Binary Gas Sorption. In Proceedings of the SPE Rocky Mountain Regional Meeting, Casper, Wyoming, 18–21 May 1992.
36. Wageningen, W.F.C.; Maas, J.G. Reservoir simulation and interpretation of the RECOPOL ECBM pilot in Poland. In Proceedings of the International Coalbed Methane Symposium, Tuscaloosa, AL, USA, 23–24 May 2007.
37. Reeves, S.; Taillefert, A. *Reservoir Modeling for the Design of the RECOPOL CO_2 Sequestration Project, Poland*; Topical Report; Advanced Resources International: Washington, DC, USA, 2002.
38. Hadro, J.; Wójcik, I. Metan pokładów węgla: Zasoby i eksploatacja. *Przegląd Geol.* **2013**, *61*, 13–15.
39. Wight, D. Unconventional drilling methods for unconventional reservoirs. *Houston Geol. Soc. Bull.* **2006**, *48*, 41.
40. Moore, T.A. Coalbed methane: A review. *Int. J. Coal Geol.* **2012**, *101*, 36–81. [CrossRef]
41. Gentzis, T.; Goodarzi, F.; Cheung, F.K.; Laggoun-Défarge, F. Coalbed methane producibility from the Mannville coals in Alberta, Canada: A comparison of two areas. *Int. J. Coal Geol.* **2008**, *74*, 237–249. [CrossRef]
42. Ham, Y.S.; Kantzas, A. Development of coalbed methane in Australia: Unique approaches and tools. In Proceedings of the CIPC/SPE Gas Technology Symposium 2008 Joint Conference, Calgary, AB, Canada, 16–19 June 2008.
43. Ren, J.; Zhang, L.; Ren, S. Multi-branched horizontal wells for coalbed methane production: Field performance and well structure analysis. *Int. J. Coal Geol.* **2014**, *131*, 52–64. [CrossRef]
44. Chen, S.; Tang, D.; Tao, S.; Yang, Y.; Chen, L. Current status and key factors for coalbed methane development with multibranched horizontal wells in the southern Qinshui basin of China. *Energy Sci. Eng.* **2019**, *7*, 1572–1587. [CrossRef]
45. Jureczka, J.; Hadro, J.; Kroplewski, Ł. Gilowice Project as a First Step to Develop Pre-Mine Drainage Technology in Poland—Update on Activities, Review of the Latest Results. XXVIII School of Underground Mining, Session–Coal Mine Methane as a Valuable Energy Source. Krakow. 27 February 2019. Available online: https://www.pgi.gov.pl/dokumenty-przegladarka/publikacje-2/biuletyn-pig/biuletyn-448/1240-biul448-1-11jureczka/file.html (accessed on 1 April 2020).
46. Polskie Górnictwo Naftowe i Gazownictwo (PGNiG). Coal-bed methane-fueled power generation unit launched by PGNiG in Gilowice. 2019. Available online: http://en.pgnig.pl/news/-/news-list/id/coal-bed-methane-fueled-power-generation-unit-launched-by-pgnig-in-gilowice/newsGroupId/1910852 (accessed on 1 April 2020).

Article

Removal of Cu (II) from Industrial Wastewater Using Mechanically Activated Serpentinite

Petros Petrounias [1,*], Aikaterini Rogkala [1], Panagiota P. Giannakopoulou [1], Paraskevi Lampropoulou [1], Petros Koutsovitis [1], Nikolaos Koukouzas [2], Nikolaos Laskaris [3], Panagiotis Pomonis [4] and Konstantin Hatzipanagiotou [1]

1 Section of Earth Materials, Department of Geology, University of Patras, 265 04 Patras, Greece; krogkala@upatras.gr (A.R.); peny_giannakopoulou@windowslive.com (P.P.G.); p.lampropoulou@upatras.gr (P.L.); pkoutsovitis@upatras.gr (P.K.); k.hatzipanagiotou@upatras.gr (K.H.)

2 Chemical Process & Energy Resources Institute, Centre for Research & Technology Hellas (CERTH), Maroussi, 15125 Athens, Greece; koukouzas@certh.gr

3 Department of Industrial Design and Production Engineering, University of West Attica, Gr-12244 Egaleo, Greece; nikolaos.laskaris@gmail.com

4 Department of Geology and Geoenvironment, National and Kapodistrian University of Athens, Panepistimioupolis Zografou, 157 84 Athens, Greece; ppomonis@geol.uoa.gr

* Correspondence: Geo.plan@outlook.com

Received: 19 February 2020; Accepted: 23 April 2020; Published: 3 May 2020

Abstract: We investigate with this study the effectiveness of mechanically activated serpentinite in capturing Cu (II) from the multi-constituent acidic wastewater of the pit lakes of the Agios Philippos mine (Greece), proposing specific areas with serpentinites suitable for such environmental applications. For this purpose ultramafic rock samples that are characterized by variable degrees of serpentinization from ophiolitic outcrops exposed in the regions of Veria-Naousa and Edessa have been examined regarding their capacity to remove the toxic load of Cu (II) from wastewater after having been mechanically activated through a Los Angeles (LA) machine (500, 1000 and 1500 revolutions). The more serpentinized and mechanically activated samples, as they have been characterized after a combination of various mineralogical, petrographic, geochemical analyses as well as after different stresses of abrasion and attrition, seem to be more effective in Cu removal than the less serpentinized ones. Selective removal of Cu (II) in the wroewolfeite phase was obtained by using the mechanically activated highly serpentinized ultramafic rocks. Furthermore, areas with highly serpentinized ultramafic rocks defined after petrographic mapping, using GIS method, which can potentially be used as filters for the effective Cu (II) removal from industrial wastewater are suggested.

Keywords: removal of Cu; mechanical activation of serpentinite; sustainability; wastewater treatment

1. Introduction

Today, societies are vastly affected by negative changes in the quality of water, air, and soil as a result of human activities. More specifically, large volumes of wastewater present within lakes, river or in industries, put intense pressure on mankind. Among the different types of pollutants, non-organic ones have a high-risk factor and are characterized as being harmful [1]. Mitigation of toxicity factors identified within water considered as waste that has been produced by industrial activities is a main concern in modern-day societies. Wastewater produced from industrial and human activities, must be brought back the societies as well as to Nature, which includes rivers, lakes and land [2]. More specifically, acidic drainage or pit lakes produced by past mining activities are held as being responsible for numerous problems associated with degraded water quality of which is responsible for limited access to water, especially for people living in surrounding regions with high poverty rates, raising risk factors for both conflicts and environmental reduction potentials for natural regeneration processes to occur [3–6]. Pit lakes are often acidic (pH 2–3), containing elevated concentrations of metals and metalloids showing a potential for the formation of highly acidic conditions, due to the fact that they exhibit relatively low buffer capacity levels [7–9]. Heavy metals are considered as high risk pollutants which are present in wastewater, including those originated from mining operations and industries which produce plants, batteries, paints and pigments, which also include industries that produce glass and ceramic materials. Wastewater often incorporates elements such as Cr, Pb, Cd, Zn, Cu and Ni [9]. When these metals are exposed to the environmental conditions, it is observed that ions are accumulated within humans, a process that will take place by consumption in the whole food chain cycle, thereby creating toxic concentrations of metals characterized as heavy that must not reach the environment [10].

Several applications have been proposed to achieve the removal of the toxic load, which include coagulation, chemical precipitation, filtration, ion exchange and solvent extraction, as well as enablement of evaporation and membrane methodologies [11]. Adsorption of heavy metals can be realized with conventional materials such as activated carbon which is often used effectively in various applications and also the use of carbon formed by organic materials (carbonized). Low cost carbon-containing materials have been used for the adsorption of inorganic pollutant materials and the removal of heavy metals, such as agricultural wastes (wool, rice, rice hulls, peat moss, cork biomass, untreated coffee grounds), modified biopolymers and industrial by-products [12,13]. Recent studies [14,15] state that peat can be regarded as highly effective in removing several heavy metals (like Hg) derived from water solutions by the performance of two types of experiments (batch and column). Peat produced by sphagnum plant moss accumulation is very effective at absorbing and removing Cd hypochlorite and oxidized $Cd(CN)_2$ plant waste, Cr^{6+} derived from aqueous solutions, and also Pb, Ni, Cu, Zn and Cd from waters considered as waste [16–19]. Biochar can be formed from various sources that include, for example: straw, shells, crop residues, wood, stover, bark, rubber, sludge, litter and peat [20–23]. Biochars produced from peat moss using pyrolysis procedures with variable carbonization conditions have raised research interest for heavy metal adsorption [24].

Numerous researchers have mentioned the use of many different rock lithotypes that include dolomite, magnesite, limestone, andesite, serpentinite and dunite not only in order to increase the pH value of industrial wastewater but also to capture heavy metals such as Cu [25,26]. Most of the aforementioned rock types are broadly used combined with other rocks as aggregate materials in a large number of construction and industrial applications while producing a vast number of sterile materials [27]. Therefore, it is important to investigate methods in order to dispose of these sterile aggregates, in accordance with the principals of sustainability, recycling, alternative use of materials, by concurrently reducing the necessity of additional energy for mineral and rock extraction. The sustainable use of natural rocks in removing heavy metals constitutes a Nature-based self-purification function, focusing on achieving "zero wastes" and "zero emissions" goals.

Serpentinite has been used as an adsorbent of Cu from industrial wastewater [26] despite the fact that the risk exists for the leaching of Cr and Ni. Its abundance in several ophiolite complexes, its wide range of construction and industrial applications combined with its particular textural characteristics make it particularly attractive for such applications.

It is widely accepted by the broad scientific community that different hydrated silicate minerals and especially clay minerals (that encompass serpentine), demonstrate abilities in the adsorption of undesirable heavy metal elements. Serpentine constitutes a group of high-Mg 1:1 layered trioctahedral minerals, with its structural formula being $Mg_6(Si_4O_{10})(OH)_8$. It is characterized as chemically simple but from the view of texture can be present in various forms. The serpentine mineral group occurs in three polymorphs: the high pressure antigorite and the low pressure chrysotile and lizardite. Lizardite in particular displays an ideal layered topology and on the other hand chrysotile has a bent structure and antigorite is often modulated [28,29]. Serpentine polymorphs belong to the phyllosilicate mineral group that displays a 1:1 layered structure, which comprises tetrahedral and octahedral sheets [28]. Four distinct O-H chain groups were identified in serpentine minerals. These are classified into two distinct topologies based on their (OH) positioning. Three of the four O-H groups are situated on the inner surface amongst two layers, whereas the third is located within the layer. The primitive unit serpentinite cell is held responsible for the cohesive attachment of the two successive layers that form hydrogen bonds [28,30].

When serpentinite is attached through the application of mechanical force activation (e.g., ball impact), this results in structural changes that mostly affect the OH groups that could potentially be brought to a looser state in the octahedral positions, enhancing the dispersal of the latter by the effects of water. When the serpentine structure is damaged through grinding, it becomes incapable of maintaining atom arrangements in a proper order to preserve its crystallinity. Scientists like Huang [31] have used activated serpentine in order to remove Cu from wastewater. The raw serpentine without grinding operation adsorbs less Cu than the chemo-mechanically-activated serpentine.

The aim of this study is the selective capture of Cu (II) within the crystalline structure of serpentine from tmechanically activated serpentinized ultramafic rocks from the acidic wastewater of a pit lake of the Agios Philippos mine (Greece). The suggestion of specific areas in Greece that present serpentinized ultramafic rocks suitable for Cu capture is a supplementary goal of this study.

2. Geological Description of Rock Materials Sources

Representative serpentinized rocks deriving from the Veria-Naousa and Edessa ophiolite complex (Greece) have been selected for studying their effectiveness in the removal of Cu from industrial wastewater after having been mechanically activated. The Veria-Naousa ophiolite belongs to the Almopias subzone (Axios geotectonic zone) in northern Greece. Its sequence consists of intense and highly tectonized serpentinized lherzolite and harzburgite of 46.76 km^2 total surface area (calculated via ArcMap 10.1), penetrated by scarce pyroxenitic dykes [32], and from gabbro, diabase and pillow basalt (Figure 1).

Remnants of oceanic lithosphere (Upper Jurassic to Late Cretaceous) constitute the Edessa ophiolite complex which was thrust out of one or more ocean basins [34–36]. This complex presents intense tectonization and consists of several tectonic units [37,38] such as serpentinized ultramafic rocks (lherzolite and harzburgite) of 50.34 km^2 total surface area (calculated via ArcMap 10.1). Moreover, diabase is the main mafic lithotype of the complex, whereas gabbro, basalt as well as diorite are less frequent (Figure 2).

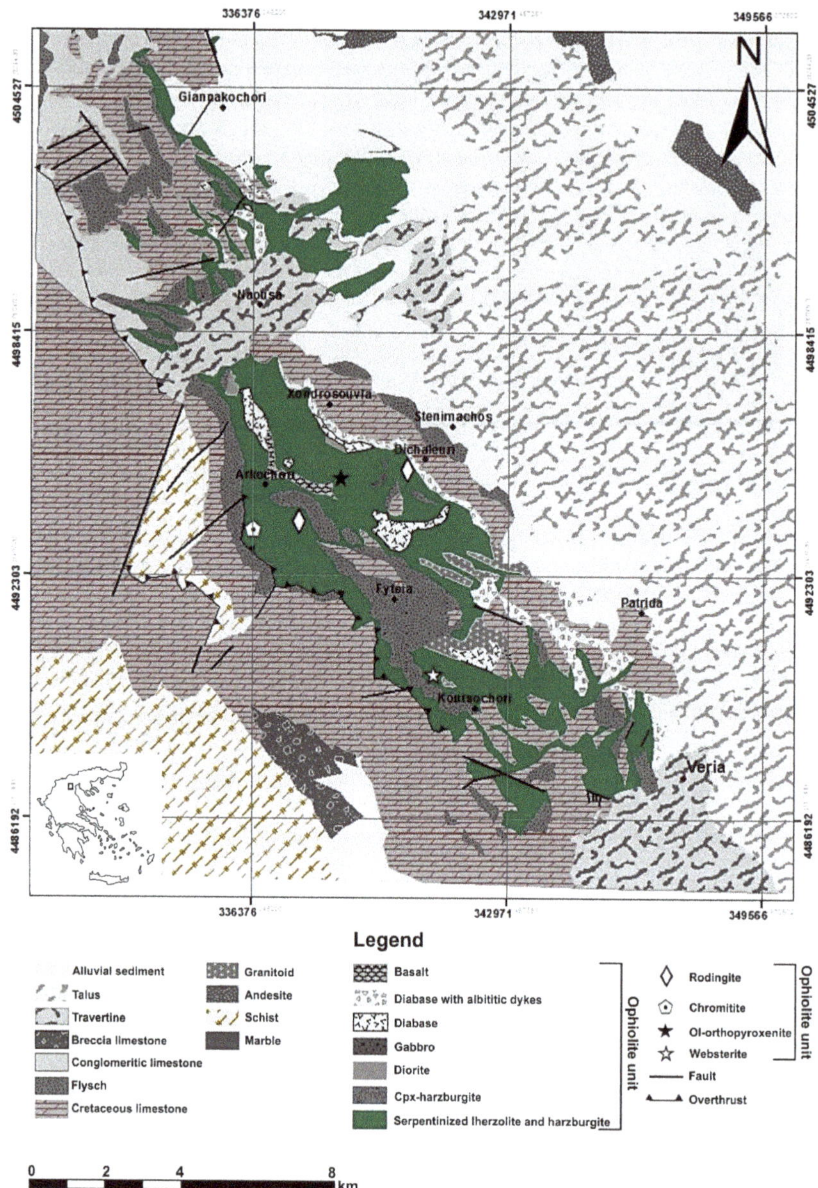

Figure 1. Modified geological map of Veria-Naousa ophiolite complex [33] after fieldwork and mapping using ArcMap 10.1; the area of investigation is shown in the black rectangle; the green color shows the serpentinized ultramafic rocks of the region.

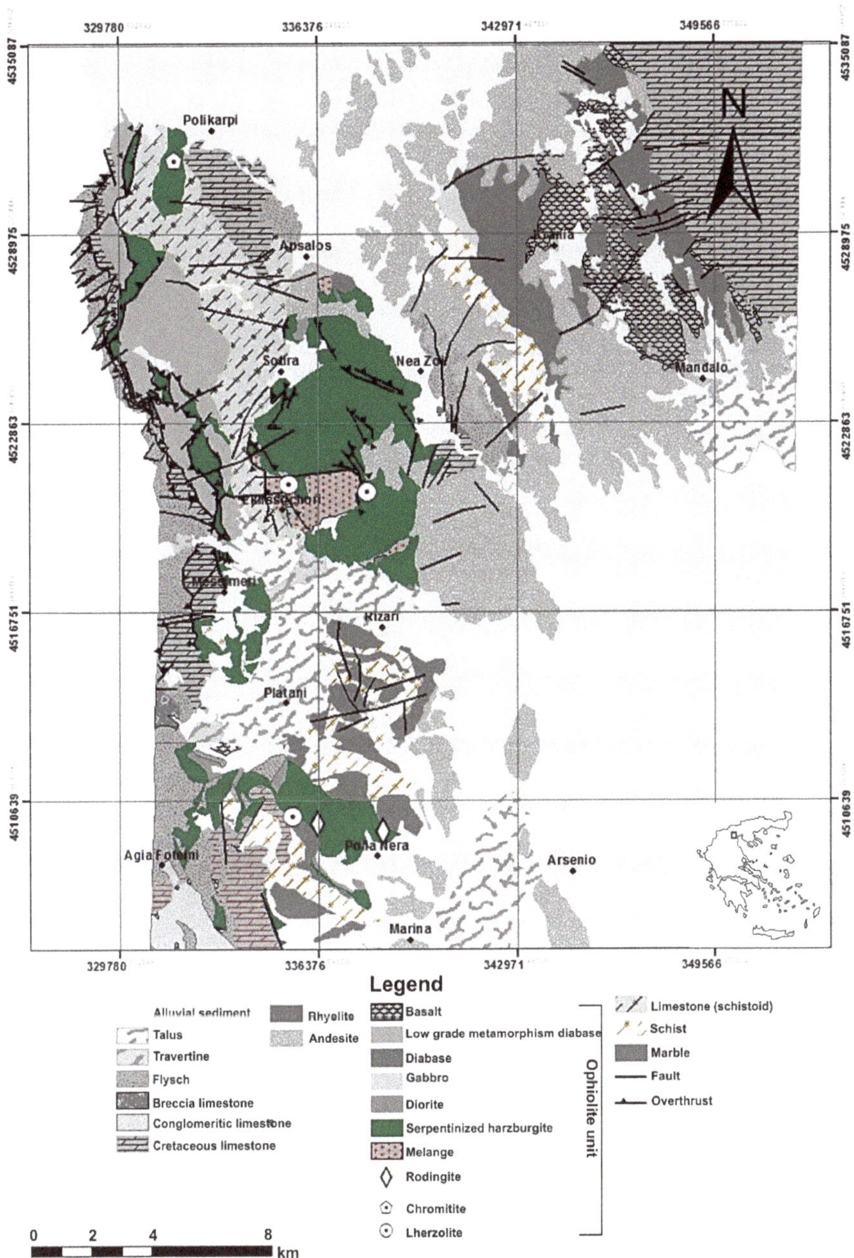

Figure 2. Modified geological map of Edessa ophiolite complex [39] after fieldwork and mapping using ArcMap 10.1; the area of investigation is shown in the black rectangle; the green color shows the serpentinized ultramafic rocks of the region.

3. Materials and Methods

Taking into account how effective the serpentinite was in removing Cu from the wastewater from the pit lake from the Agios Phillippos Kirkis mine [26], four serpentinites, derived from the aforementioned ophiolite complexes and presenting different degree of serpentinization were used after their mechanical activation to various degrees to examine the mechanism through which serpentinite can become more effective in the removal of Cu.

3.1. Characterization of the Studied Rock Materials

The first stage for the characterization of the investigated rocks is their petrographic examination (mineralogical and textural features) with the aid of a Leitz polarizing optical microscope (Leica Microsystems, Wetzlar, Germany). The modal composition of the investigated serpentinites was also identified using powder X-ray diffraction (XRPD) analysis, with a Bruker D8 Advance Diffractometer (Bruker, Billerica, MA, USA), with Ni-filtered CuKα radiation. The scanning area for bulk mineralogy of the samples covered the 2θ interval 2–70°, with a scanning angle step size of 0.015° and a time step of 0.1 s. The modal composition was identified with the DIFFRACplus EVA 12® software (Bruker-AXS) based on the ICDD Powder Diffraction File of PDF-2 2006. The mineral phases were semi-quantified with the TOPAS 3.0 ® software (TOPAS MC Inc., Oakland, CA, USA), based on the Rietveld method refinement routine. The routine is based on the calculation of a single mineral-phase pattern regarding to the crystalline structure of each mineral, and the refinement of the pattern using a non-linear least squares routine. According to the study by Bish and Post [40], the quantification errors were calculated and they are estimated to be ~1%. Polished thin sections of the examined rock samples were analyzed in scanning electron microscopes (SEM) in order to identify their mineralogical characteristics. Serpentine minerals microanalyses were carried out in a JEOL JSM-6300 SEM equipped with energy dispersive (EDS) using the INCA software (version). The scanning electron microscope used is located in the Laboratory of Electron Microscopy and Microanalysis (University of Patras, Greece). Operating conditions were accelerating voltage 25 kV and beam current 3.3 nA, with a 4 μm beam diameter. The total counting time was 60 s and dead-time 40%. Synthetic oxides and natural minerals were used as standards for the analyses, where the detection limits are ~0.1% and accuracy better than 5% was obtained. Furthermore, an XRF (X-Ray fluorescence) spectrometer and a sequential spectrometer (ICP-ES) were used for the determination of the major and trace elements of the studied samples, which were carried out at Bureau Veritas Mineral Laboratories (Vancouver, BC, Canada).

3.2. Methodology for the Cu (II) Removal

The studied serpentinites were placed in the Los Angeles (LA) machine in order to become mechanically activated after they have been sieved in the No. 14 sieve (1.40 mm). The Los Angeles (LA) machine is a rotating drum which contains certain number of steel spheres. Los Angeles (LA) test constitutes a basic test which indicates the mechanical quality of aggregates rocks by identifying their resistance in abrasion and attrition. At this stage, serpentinite samples were rotated in the LA machine for 500, 1000 and 1500 revolutions, respectively. Then, the samples were sieved appropriately in order to reach the grain size of 0.8–0.6 mm which then used as filters in batch type columns. Serpentinites that were not mechanically activated were also used as filters in the batch type columns after having been sieved appropriately in order to reach a grain size between 0.8 and 0.6 mm.

The next stage included an experimental arrangement consisting of columns of borosilicate glass of 20 mm diameter. In each column 70 gr of the different mechanically activated serpentinites (derived from the different revolutions used in the LA machine) was used, and 300 mL of wastewater was passed through each column four times. After this four pass procedure, wastewater samples were further processed for geochemical analysis of Cu concentration, which was performed in the Institute for Solid fuels Technology and Application of the National Centre for Research and Technology Hellas using conjugated plasma argon mass spectrometry. All the sorption experiments were carried out at room temperature (25 ± 2 °C).

4. Results

4.1. Results of the Studied Rock Materials

4.1.1. Petrographic Features of the Studied Rock Materials

Two types of serpentinized ultramafic rocks regarding their degree of serpentinization were collected from the studied areas. Their textural and mineralogical characteristics are analyzed below in Figure 3.

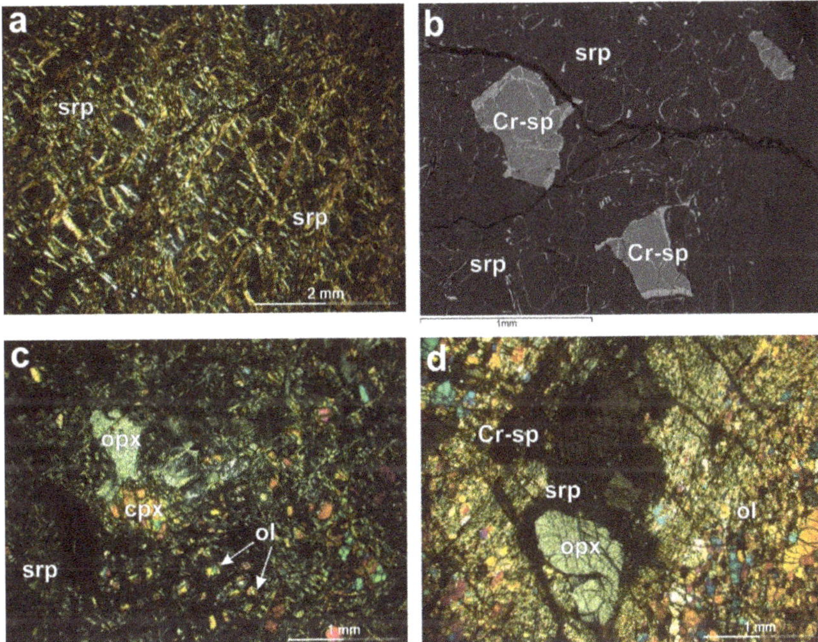

Figure 3. Textural features of serpentinites from the investigated ophiolite complexes: (**a**) Photomicrograph of mesh texture in serpentinized (srp) matrix (sample BE.01, + Nicols); (**b**) Back-scattered electron image (BSE) of Cr-spinel (Cr-sp) with curved boundaries and thin rims of ferritchromit and magnetite as well as mesh serpentine (srp) (sample ED.111); (**c**) Photomicrograph of an orthopyroxene (opx) porphyroclast and clinopyroxene porphyroclast (cpx) surrounded by olivine neoblasts (ol) in serpentinized (srp) matrix (sample BE.118, + Nicols); (**d**) Photomicrograph of orthopyroxene porphyroclast (opx) and olivine grains (ol), subhedral Cr-spinel (Cr-sp) and ribbon serpentine (srp) (sample ED.86B, + Nicols).

(1) Highly-serpentinized (Group I)

The primary modal mineralogical assemblage of the highly serpentinized ultramafic rocks is intensely altered, whereas relics of orthopyroxene and Cr-spinel were observed. The subhedral to euhedral Cr-spinel are frequently crosscut by secondary veins and/or surrounded by thin rims of magnetite and ferritchromit because of the effects of ocean floor alteration processes. Serpentine is the main alteration phase which forms mesh, ribbon, bastite and hourglass textures. Moreover, less chlorite and magnetite subsequently observe, likely being the result of retrograde metamorphism during exhumation (Figure 3).

(2) Medium-Serpentinized (Group II)

The primary assemblage of these serpentinized ultramafic rocks comprises of orthopyroxene, olivine, less of clinopyroxene, olivine and Cr-spinel constituting less than 30% of the whole assemblage. Porphyroclasts of orthopyroxene exhibit exsolution lamellae of clinopyroxene, typical characteristic of mantle peridotites. Olivine displays porphyroclastic grains and smaller neoblasts. Locally, olivine porphyroclasts show strain lamellae, undulose extinction, shearing and recrystallization (Figure 3). Clinopyroxene appears as relict subhedral porphyroclasts. Crystals of Cr-spinel presented as subhedral to euhedral and they display an irregular distribution of ferritchromit compositional areas. The boundary between the Cr-spinel (either unaltered or altered) and the ferritchromit is curved and lobate. The main secondary product is serpentine which predominantly displays ribbon and mesh textures, whereas others are chlorite and magnetite. The modal composition of the investigated serpentinites was further determined by XRPD analysis in Figure 4.

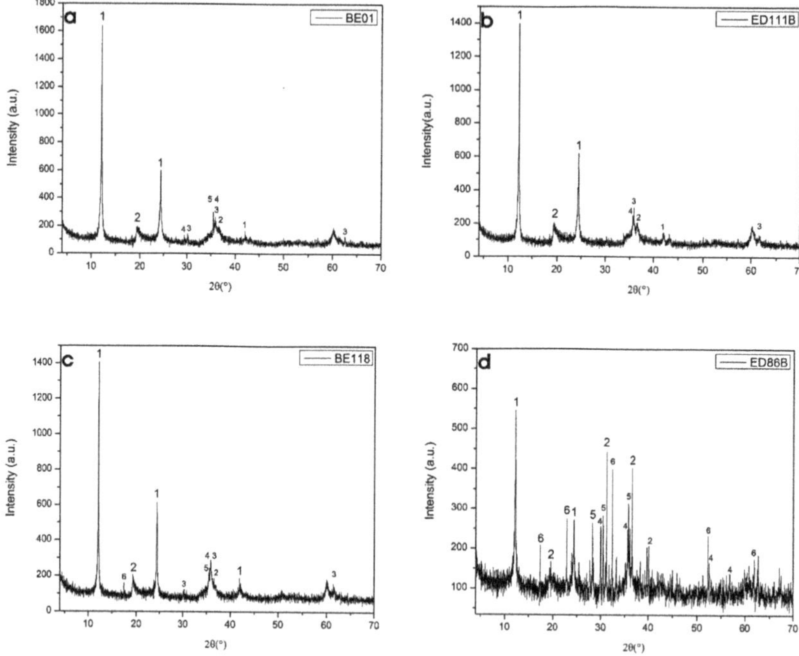

Figure 4. XRPD patterns of the examined serpentinites: (**a**) highly serpentinized ultramafic rock (sample: BE.01); (**b**) highly serpentinized ultramafic rock (sample: ED.111B); (**c**) medium serpentinized ultramafic rock (sample: BE.118); (**d**) medium serpentinized ultramafic rock (sample: ED.86B), (1: Serpentine, 2: Spinel, 3: Magnetite, 4: Clinopyroxene, 5: Orthopyroxene, 6: Olivine).

4.1.2. XRPD of Mineral Rock Materials

The X-ray diffraction enabled us to identify the crystalline phases of the tested ultramafic rocks where higher picks of serpentine were observed in samples BE.01 and ED.111B whereas samples BE.118 and ED.86B display lower picks of serpentine (Figure 4). Serpentine content as well the content of other mineralogical phases of the studied rock samples were calculated via Rietveld method and are listed in the Table 1, where the sample BE.01 presents as the most serpentinized, while ED.86B presents as the less serpentinized sample. The identification of the degree of serpentinization of the studied rock samples through the petrographic analysis via polarizing microscope is in accordance with the results of Rietveld method.

Table 1. Semi-quantitative mineralogical assemblage of the tested serpentinites. The quantification errors calculated for each phase according to Bish and Post [40] are estimated to be ~1% (ol: olivine, opx: orthopyroxene, cpx: clinopyroxene, sp: spinel, mgt: magnetite, srp: serpentine, BE: Veria-Naousa ophiolite, ED: Edessa ophiolite) (-: below detection limit).

Samples	ol	opx	cpx	sp	mgt	srp
BE.01	-	2.3	1.7	3.3	1.0	91.7
ED.111B	-	-	1.1	7.1	3.2	88.6
BE.118	3.0	14.0	7.0	4.5	1.5	70.0
ED.86B	11.5	15.5	11.0	3.0	-	59.0

4.1.3. Chemistry of Serpentine Minerals

Representative microanalyses of the serpentine minerals are shown in Table 2 and plotted in the diagrams of Figure 5. Serpentine minerals from Veria-Naousa and Edessa ophiolite are composed of SiO_2 (42.02–46.27 wt.%), MgO (36.45–41.50 wt.%), Fe_2O_3 (1.10–6.53 wt.%) and less Al_2O_3, CaO and TiO_2. Serpentine minerals from Group I display higher SiO_2, MgO, Fe_2O_3 and lower Al_2O_3, CaO and TiO_2 contents than those of Group II. NiO and Cr_2O_3 contents are wide (0.00–2.20 wt.% and 0.00–1.93 wt.%, respectively). This wide range of NiO and Cr_2O_3 content may be connected with the nature of the replaced olivine in the case of Ni-rich serpentine and pyroxene as well as in the case of Cr-rich serpentine.

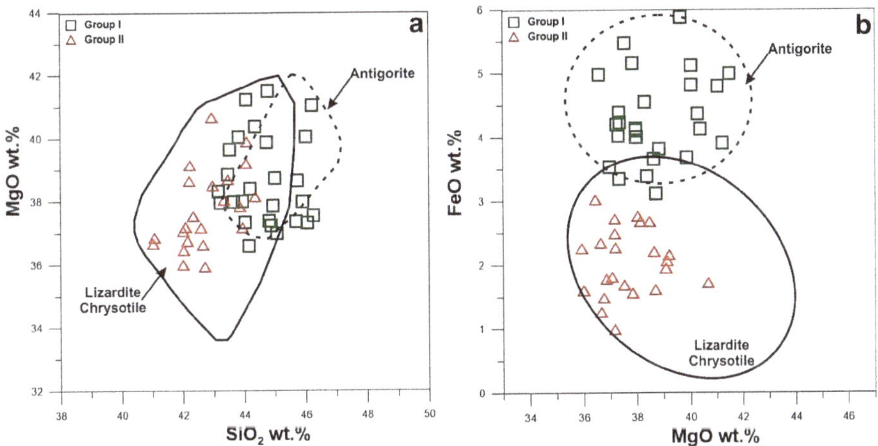

Figure 5. (**a**) MgO vs. SiO_2 plot and (**b**) FeO vs. MgO plot for the analyzed serpentine minerals in serpentinites from Veria-Naousa and Edessa ophiolites. Fields of lizardite, chrysotile and antigorite are from Singh & Singh [41].

Table 2. Representative electron microanalyses of serpentine minerals from serpentinites of Veria-Naousa and Edessa ophiolites (-: below detection limit).

	Group I								Group II							
Sample	BE.01				ED.111B				BE.118				ED.86B			
Anal. No	14	18	21	23	3	7	10	14	7	13	17	20	4	5	8	10
wt.%																
SiO_2	46.27	44.16	44.78	44.94	43.22	43.53	44.37	45.01	42.95	44.37	42.04	42.59	43.34	42.02	43.94	43.86
TiO_2	-	-	-	-	-	-	-	-	-	0.07	0.09	0.05	-	-	-	-
Al_2O_3	0.43	-	-	0.22	-	-	-	-	-	-	0.66	0.96	-	0.70	1.67	2.58
Fe_2O_3	6.08	5.53	5.55	5.73	4.45	6.53	4.58	3.46	1.91	2.98	3.36	3.02	3.07	2.01	1.10	1.73
MnO	-	-	-	-	-	-	-	-	-	0.06	-	-	-	-	-	-
MgO	37.55	36.58	41.50	37.85	37.96	39.65	40.38	38.72	40.66	38.14	36.45	37.17	38.05	37.06	37.17	37.84
CaO	-	-	-	-	-	-	-	-	0.18	0.08	-	0.14	-	-	0.08	0.19
Na_2O	-	-	-	-	-	-	-	-	-	-	-	0.06	-	-	-	-
K_2O	-	-	-	-	-	-	-	-	-	-	-	-	-	-	-	-
NiO	-	-	-	-	-	-	-	-	-	-	-	-	2.20	-	0.39	0.45
Cr_2O_3	0.39	-	-	-	0.81	-	-	-	-	0.05	-	-	-	0.64	1.35	1.93
Sum	90.72	86.27	91.83	88.74	86.44	89.71	89.33	87.19	85.70	85.75	82.60	83.99	86.66	82.43	85.70	88.58
Formula units based on 7 atoms of oxygens																
Si	2.068	2.073	1.988	2.054	2.030	1.985	2.015	2.078	2.020	2.081	2.051	2.042	2.041	2.049	2.058	2.000
Ti	-	-	-	-	-	-	-	-	-	0.002	0.003	0.002	-	-	-	-
Al	0.023	-	-	0.012	-	-	-	-	-	-	0.038	0.054	-	0.040	0.092	0.139
Fe^{3+}	0.205	0.195	0.185	0.197	0.157	0.224	0.157	0.120	0.068	0.105	0.123	0.109	0.109	0.074	0.039	0.059
Mn	-	-	-	-	-	-	-	-	-	0.002	-	-	-	-	-	-
Mg	2.502	2.560	2.746	2.579	2.658	2.695	2.734	2.664	2.850	2.666	2.650	2.657	2.671	2.694	2.595	2.572
Ca	-	-	-	-	-	-	-	-	0.009	0.004	-	0.007	-	-	0.004	0.009
Na	-	-	-	-	-	-	-	-	-	-	-	0.006	-	-	-	-
K	-	-	-	-	-	-	-	-	-	-	-	-	-	-	-	-
Ni	-	-	-	-	-	-	-	-	-	-	-	-	0.083	-	0.015	0.017
Cr	0.014	-	-	-	0.030	-	-	-	-	0.002	-	-	-	0.025	0.050	0.070
Total	4.811	4.829	4.919	4.842	4.876	4.903	4.906	4.862	4.947	4.863	4.866	4.877	4.905	4.882	4.852	4.866

In the diagram of Figure 5a, the analyzed serpentine minerals from Group I are mostly antigorite, while Group II are mostly lizardite and chrysotile and less antigorite. The similarity in characteristics is consistent with the plot of MgO vs. FeO (Figure 5b). FeO was calculated by the Fe_2O_3 method in order produce the diagram shown in Figure 5b.

4.1.4. Geochemical Features of Rock Materials

Major and trace elements data from the studied serpentinites, along with their loss on ignition (LOI) values are listed in Table 3. LOI values vary significantly from 7.9% to 14.6% with the most serpentinized sample (BE.01) presenting the higher LOI value, while the least serpentinized one (ED.86B) presenting the lowest LOI value.

Table 3. Representative geochemical analyses of the studied serpentinites (-: below detection limit, total).

	Group I		**Group II**	
Sample	**BE.01**	**ED.111B**	**BE.118**	**ED.86B**
	Major elements (wt.%)			
SiO_2	39.82	40.21	40.17	42.01
TiO_2	-	-	0.01	0.13
Al_2O_3	1.01	0.44	1.48	3.57
Fe_2O_3	8.86	8.89	7.43	8.06
MnO	0.11	0.06	0.12	0.11
MgO	34.17	33.54	36.28	34.17
CaO	0.10	0.16	1.46	3.17
Na_2O	-	-	-	0.18
K_2O	-	-	-	-
P_2O_5	-	-	0.01	0.02
LOI	14.6	14.0	12.4	7.9
Total	98.67	99.30	99.36	99.32
	Trace elements (ppm)			
Cr	2963	2333	2292	2607
Co	91.1	88.4	100.7	92.9
Ni	2655.8	2958.7	2185.0	1812.2
Cu	12.9	8.2	5.8	5.1
Zn	8	30	24	15
Pb	21.7	0.8	0.6	0.3
As	7.2	2.6	-	-
U	0.2	3.0	-	-

4.2. *Experimental Study Results*

4.2.1. Chemical Analysis of the Wastewater

The wastewater from the selected pit lake of the Agios Philippos mine is characterized by an acidic pH value (2.99) and was geochemically analyzed by Petrounias et al. [26]. These results are given in Table 4. According to the results of Table 4, the wastewater contains an extremely high concentration of Cu (8847.21 ppb).

Table 4. Chemical analysis of wastewater (-: below detection limit) [26].

Elements (ppb)	Concentrations
Ag	0.18
As	-
Ba	20.54
Be	24.11
Cd	1717.73
Co	213.11
Cr	-
Cs	13.70
Cu	8847.21
Ga	3.90
Li	25.96
Mn	70,982.00
Pb	812.77
Rb	60.42
Sr	520.57
V	-
U	111.96
Zn	285,458.55
Se	24.91
Ni	1149.40
Fe	6149.02

4.2.2. Chemical Analysis of Wastewater after Having Penetrated 4 Times through Columns of Batch Type

Table 5 displays the final results of the chemical analyses of wastewater after having been passed four times through each column, in which the different mechanically activated serpentinites were used as filters. As displayed in Table 5, high amounts of Cu have been trapped from the samples that contain higher serpentinite contents.

Table 5. Chemical analysis of the wastewater.

	Cu (ppb)			
Samples/Revolutions	0	500	1000	1500
BE.01	4563.20	80.35	65.40	25.21
ED.111B	6380.51	121.30	85.47	35.30
BE.118	7500.46	3670.40	1670.40	320.48
ED.86B	8070.31	6120.00	2720.50	670.12

More specifically, the highly serpentinized sample (BE.01) seems to present a higher Cu removal capacity in comparison to the other three rock samples, even when used without having been mechanically activated, as well as when it was mechanically activated after 500, 1000 and 1500 revolutions, respectively.

4.2.3. Chemical X-ray Diffractometry of Serpentinites after the Experimental Study

The mechanically activated serpentinites (after 1500 revolutions), after having been used as filters in the experimental arrangement of the batch type columns for the removal of Cu, were analyzed by X-ray diffractometry. The corresponding X-ray diffraction patterns are given in Figure 6. In the pattern of samples BE.01 and of ED.111B, which are the most serpentinized, the mineral wroewolfeite [$Cu_4(SO_4)(OH)_6 2H_2O$] appears, in contrast to the other two samples which are characterized by lower serpentinite contents (samples BE.118 and ED.86B).

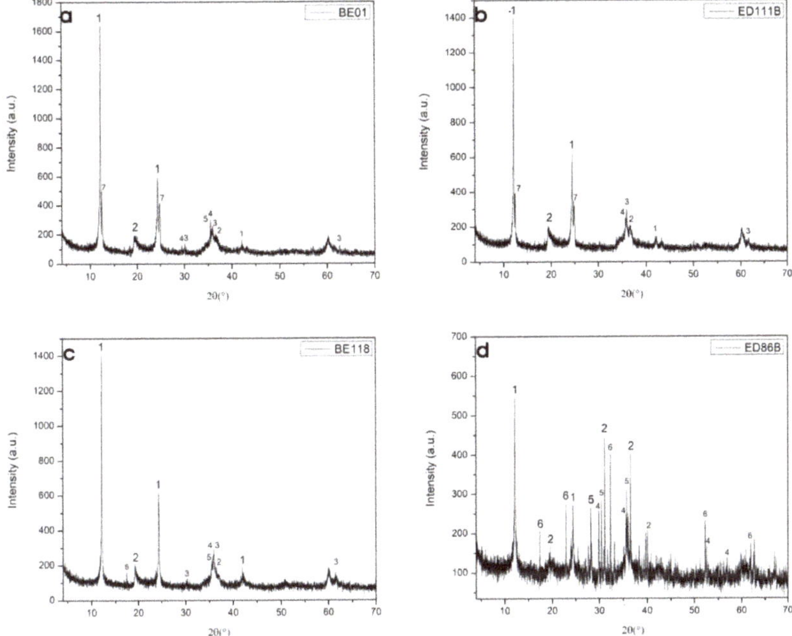

Figure 6. XRPD patterns of the investigated serpentinites: (**a**) highly serpentinized ultramafic rock (sample: BE.01); (**b**) highly serpentinized ultramafic rock (sample: ED.111B); (**c**) medium serpentinized ultramafic rock (sample: BE.118); (**d**) medium serpentinized ultramafic rock (sample: ED.86B), (1: Serpentine, 2: Spinel, 3: Magnetite, 4: Clinopyroxene, 5: Orthopyroxene, 6: Olivine, 7: Wroewolfeite).

5. Discussion

Several researchers have studied the individual use of minerals and of rocks materials in various environmental applications and more specifically for the removal of toxic loads using these materials [25,26]. Furthermore, quite many researchers have carried out mechanical, chemical and mechanochemical activation of minerals and rocks in order to increase their effectiveness in removing toxic loads from wastewater. Huang et al. [31] have concluded that mechanochemically activated serpentine presents very satisfactory results concerning Cu (II) removal (to an almost complete degree), allowing this type of removal to act as a way/method of recycling and reusing Cu derived from various wastewater solutions. Mechanical activation constitutes a way of applying mechanical stress to induce changes in the surface properties as well as in the crystalline structure of minerals [42–45]. Even though numerous studies relative to the understanding of the nature of the effect of mechanical loads on structural changes in the structure of minerals have been carried out, the activation of low cost phyllosilicate minerals, such as serpentine, has not yet been extensively investigated in environmental metal neutralization uses.

Petrounias et al. [26] combined sterile raw materials by using sterile aggregates of the LA test, and achieved significant removal of Cu from the Agios Philippos Kirkis mine (Greece). They attributed the Cu removal to the existence of mechanically activated serpentinite. This was carried out under pH 4 conditions, where precipitation of metal complexes is not favored [46]. In general, the pH of the solution controls the adsorption of Cu in the laminated serpentine, mainly of tetrahedral silicate during the ion exchange process.

This study, which is based on and expands the initial study of Petrounias et al. [26] attempted to find out, more accurately, the mechanism as well as the necessary and sufficient conditions for a serpentinized ultramafic rock to work efficiently in selective Cu capture from wastewater. As presented in Table 5, it is obvious that as the mechanical stress increases (through the LA test), which means that as the loadings of abrasion and attrition increase, the capacity of the serpentinized rocks in capture Cu increases regardless their degree of serpentinization. Thus, this increase may happen exclusively due to the existence and the structure of serpentine. When serpentine is subjected to mechanical loading there is structural change, especially in the OH^- group, which may become more relaxed than in the octahedral positions and easily circulate in solution when in contact with it [31]. When the structure of serpentine is decomposed by the use of abrasive mechanical stresses, is is unable to retain its atom structure and thus its crystallinity is altered. This change in the crystallinity of serpentine has been cited as one of the major factors for copper adsorption [26,31]. It is understood that mechanical stress may rupture the tetrahedral and octahedral sheets of serpentine. Moreover, intensively significant index relative to the capacity of Cu capture seems to be the degree of serpentinization of the rock materials used. More specifically, the more serpentinized ultramafic rocks (BE.01, ED.111B), as they have been identified through the petrographic observation (Figure 3), the X-Ray diffractometry and Rietveld method (Figure 4, Table 1) and the geochemical analysis given through the LOI index (Table 3), seem more effective in Cu capture (Table 5) relative to those contained less serpentine (BE.118, ED.86B). Moreover, the effectiveness, in the Cu capture, of the most serpentinized ultramafic rocks is indicated by the presence of wroewolfeite in the structure of these rocks after their use as filters (Figure 6).

The amount of serpentine found in the studied rocks is presented as the determinant factor, as through this, the rate of change of available crystalline meshes capable of Cu capture from the studied solutions is determined. However, the combination of the abovementioned factors seems to constitute the more crucial combination concerning the effectiveness of each serpentinized lithotypes for the removal of Cu. More specifically, the highly serpentinized ultramafic rocks (BE.01, ED.111B with 88.6–91.7% of serpentine contained) present faster and more efficient capture of Cu in contrast to those characterized by lesser amounts of contained serpentine (BE.118, ED.86B with 59.0–70.0% of serpentine).

Furthermore, the Cu capture capacity seems to be related to the type of serpentine and therefore to the special texture characteristics of the studied serpentinites. More specifically, the serpentinized samples of similar degree of serpentinization (Table 1) but with a variety in textural features (mesh, ribbon) and in serpentine type (chrysotile, antigorite and lizardite), as shown in the corresponding diagrams (Figure 5), display significant differences regarding their Cu capture effectiveness. Serpentine from highly serpentinized ultramafic rocks (Group I) plotted in the field of antigorite (Figure 5) present as more capable of removing Cu in contrast those of Group II (medium serpentinized ultramafic rocks) whose serpentine is plotted in the field of lizardite and chrysotile. This may happen due to the different capacity of lizardite, chrysotile and antigorite for retaining their crystallinity under mechanical stress. Additionally, the petrographic characteristics such as texture may significantly influence the Cu capture. More specifically, the mesh texture of serpentine in the most serpentinized sample (BE.01) is presented as the most effective texture than the other textures of the rest studied rocks, a fact which may happen due to surface adsorption of Cu^{2+} when $CuOH^-$ is simultaneously generated and adsorbed within the microcells.

Proposed Areas with Serpentinites for Their Potential Use as Filters for Cu Removal

Nowadays, in the era of cyclical economy and climate change, the ability to recover and reuse metals has been studied by many researchers [31,47]. Such research is fundamental to modern cutting-edge studies. In this study, the amount of Cu captured using serpentinites from the Veria-Naousa and Edessa ophiolite complexes suggest promising results to potentially recover copper accumulated within the crystallinity of serpentine (Figure 6), by using a variety of physicochemical recovery methods. The main object of the this work is to capture Cu within the crystallinity of serpentine from serpentinized ultramafic rocks and also to examine this application in conjunction with the recovery of copper from wastewater, as well as to propose potential areas that encompass serpentinized ultramafic rocks suitable for these applications (Figure 7).

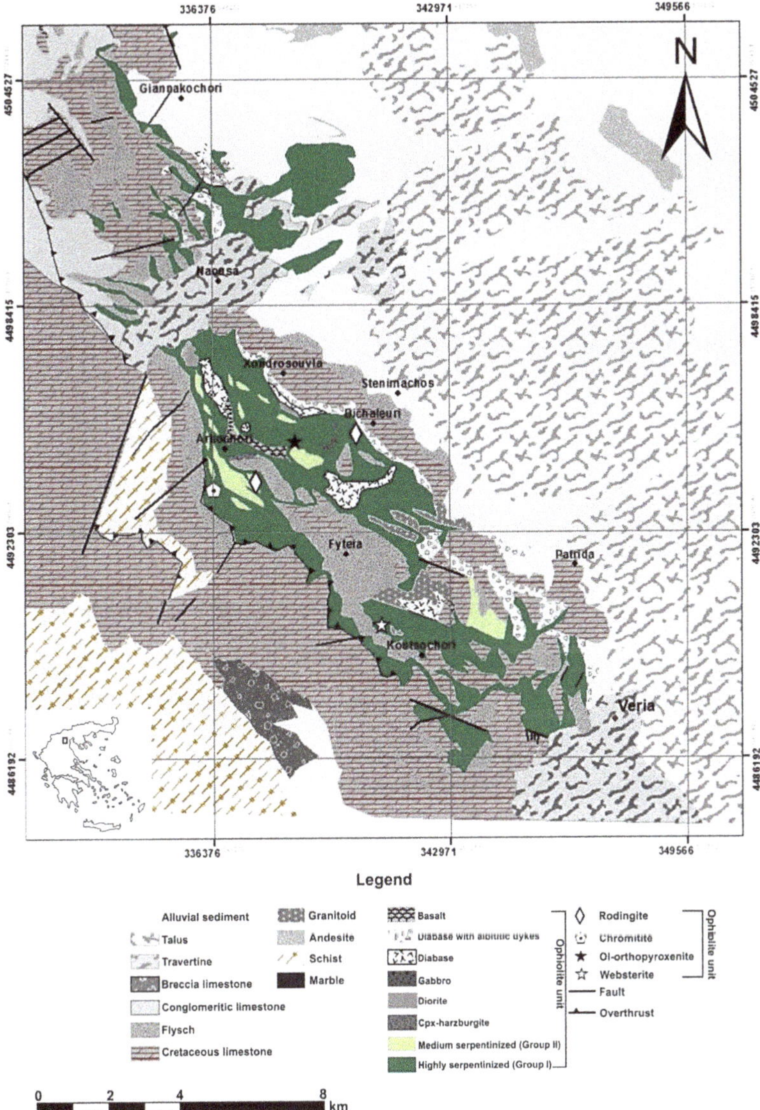

Figure 7. Modified geological map of Veria-Naousa ophiolite complex [33] after fieldwork and mapping using ArcMap 10.1; the area of investigation is shown in the black rectangle; the dark green color shows the highly-serpentinized ultramafic rocks of the region and the light green color shows the medium-serpentinized ultramafic rocks.

GIS-based petrographic mapping of serpentinites feasible to be used as raw materials for copper capture are suggested. In the modified maps (see below) we have taken into account all of the aforementioned factors and propose specific areas that are defined and measured from the examined ophiolite complexes (Figures 7 and 8). Specifically, the areas marked with dark green color (zone

A) constitute the suggested areas from the mentioned complexes with the suitable serpentinites. These serpentinites can potentially be used as filters for Cu removal regarding their petrographic characteristics (degree of serpentinization, textural features and serpentine type) and geochemical characteristics (LOI index). The Veria-Naousa ophiolite complex comprises of highly-serpentinized ultramafic rocks, and GIS-based calculations yield an area of 43.84 km^2, whereas the Edessa ophiolite complex is calculated to be 42.19 km^2 (Figure 7). On the other hand, zone B, marked with light green color, constitutes the lesser suitable areas for Cu capture, as they contain lower amounts of serpentine and different mineralogical and petrographic characteristics than those of zone A.

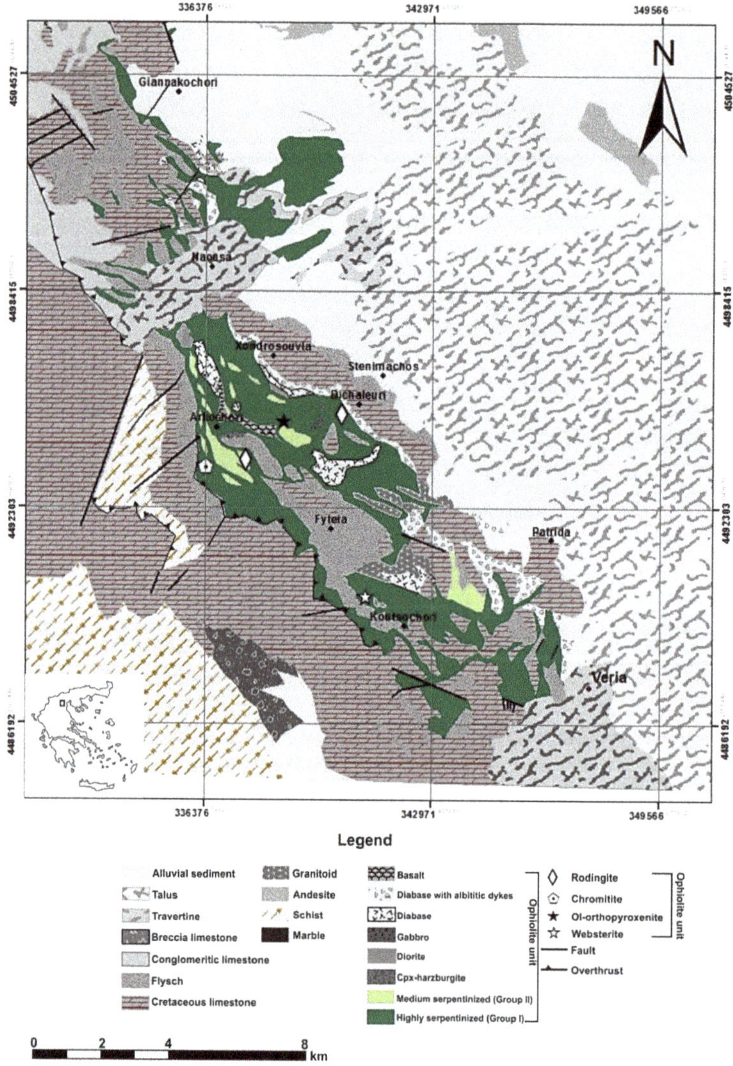

Figure 8. Modified geological map of Edessa ophiolite complex [39] after fieldwork and mapping using ArcMap 10.1; the area of investigation is shown in the black rectangle; the dark green color shows the highly-serpentinized ultramafic rocks of the region and the light green color shows the medium-serpentinized ultramafic rocks.

Concerning the Veria-Naousa ophiolite complex, 2.93 km^2 of highly serpentinized ultramafic rocks was calculated using GIS method, whereas in the Edessa ophiolite complex 8.15 km^2 was calculated (Figure 8). Conclusively, the highly serpentinized ultramafic rocks from both ophiolite complexes are present in higher volumes than those of the medium-serpentinized rocks. Additionally, through the proposed maps, which are first introduced through this study, an extra use for these serpentinites is proposed apart from aggregates for construction and environmental applications as has already highlighted by Petrounias et al. [48,49].

6. Conclusions

In this paper, the selective capture of Cu (II) from acidic wastewater derived from the pit lakes of the Agios Philippos mine (Greece) from mechanically activated serpentinite using a LA machine was studied. The abovementioned study leads to the following conclusions:

- Two groups of serpentinized ultramafic rocks from the investigated ophiolite complexes which were detected according to their petrographic observations were in accordance with their mineralogical, chemical and geochemical analyses.
- The highly serpentinized rocks (Group I) are more effective in Cu removal in contrast to the medium serpentinized rock samples (Group II).
- The higher mechanical activation of the studied serpentinites (after they have been subjected to a 1500 revolutions LA test) is related to their higher capability to perform Cu removal.
- Selective removal of Cu (II) in the form of the wroewolfeite phase was achieved by using mechanically activated highly serpentinized ultramafic rocks.

Furthermore, areas with highly serpentinized ultramafic rocks that can potentially be used as filters for the effective Cu (II) removal from industrial wastewater are suggested and more specifically these are the 43.84 km^2 areas from the Veria-Naousa ophiolite and the 42.19 km^2 area from the Edessa ophiolite.

Author Contributions: P.P. (Petros Petrounias) participated in the fieldwork, the elaboration of laboratory tests, the interpretation of the results, coordinated the research and the writing of the manuscript; A.R. participated in the fieldwork, performed the SEM work, the interpretation of the results and contributed to the manuscript writing; P.P.G. participated in the elaboration of laboratory tests, the interpretation of the results and contributed to the manuscript writing; P.L. carried out the XRPD analyses, participated in the interpretation of the results and contributed to the manuscript writing; P.K. participated in the elaboration of laboratory tests and in the interpretation of the results; N.K. contributed to the manuscript writing; N.L. participated in the fieldwork and in the interpretation of the results; P.P. (Panagiotis Pomonis) contributed to the manuscript writing; K.H. participated in the interpretation of the results. All authors have read and agreed to the published version of the manuscript.

Funding: This research received no external funding.

Acknowledgments: We would like to thank A.K. Seferlis of the Laboratory of Electron Microscopy and Microanalysis, University of Patras for his aid, E. Gianni for her assistance in the preparation of the studied samples for laboratory tests. We also thank M. Kalpogiannaki for her assistance in the construction of the geological maps.

Conflicts of Interest: The authors declare no conflict of interest.

References

1. Ms, A.-R.; Arm, A.-R. Removal of Heavy Metals from Industrial Waste Water by Biomass-Based Materials: A Review. *J. Pollut. Eff. Control.* **2016**, *5*. [CrossRef]
2. Weber, W.J.; McGinley, P.M.; Katz, L.E. Sorption phenomena in subsurface systems: Concepts, models and effects on contaminant fate and transport. *Water Res.* **1991**, *25*, 499–528. [CrossRef]
3. Salmon, S.U.; Oldham, C.E.; Ivey, G.; Salmon, U. Assessing internal and external controls on lake water quality: Limitations on organic carbon-driven alkalinity generation in acidic pit lakes. *Water Resour. Res.* **2008**, *44*, 10414. [CrossRef]
4. Schindler, D.W. The significance of in-lake production of alkalinity. *Water Air Soil Pollut.* **1986**, *30*, 931–944. [CrossRef]

5. Geller, W.; Klapper, H.; Salomons, W. *Acidic Mining Lakes*; Springer: Berlin/Heidelberg, Germany, 1998.
6. Geller, W.; Koschorreck, M.; Wendt-Potthoff, K.; Bozau, E.; Herzsprung, P.; Büttner, O.; Schultze, M. A pilot-scale field experiment for the microbial neutralization of a holomictic acidic pit lake. *J. Geochem. Explor.* **2009**, *100*, 153–159. [CrossRef]
7. Lattanzi, P.; Da Pelo, S.; Musu, E.; Atzei, D.; Elsener, B.; Fantauzzi, M.; Rossi, A. Enargite oxidation: A review. *Earth-Sci. Rev.* **2008**, *86*, 62–88. [CrossRef]
8. Sperling, E.; Grandschamp, C.A.P. Possible water uses in mining lakes: Case study of Agua Claras, Brazil. In Proceedings of the 33rd WEDC International Conference, Accra, Ghana, 7–11 April 2008; WEDC, Loughborough University: Loughborough, UK, 2008; pp. 375–380.
9. Shevenell, L.; A Connors, K.; Henry, C.D. Controls on pit lake water quality at sixteen open-pit mines in Nevada. *Appl. Geochem.* **1999**, *14*, 669–687. [CrossRef]
10. Meena, A.K.; Kadirvelu, K.; Mishra, G.; Rajagopal, C.; Nagar, P. Adsorptive removal of heavy metals from aqueous solution by treated sawdust (Acacia arabica). *J. Hazard. Mater.* **2008**, *150*, 604–611. [CrossRef]
11. Panayotova, M.; Velikov, B. Influence of zeolite transformation in a homoionic form on the removal of some heavy metal ions from wastewater. *J. Environ. Sci. Health Part A* **2003**, *38*, 545–554. [CrossRef]
12. Sekhar, K.C.; Kamala, C.; Chary, N.; Anjaneyulu, Y. Removal of heavy metals using a plant biomass with reference to environmental control. *Int. J. Miner. Process.* **2003**, *68*, 37–45. [CrossRef]
13. Babel, S.; Kurniawan, T.A. Various treatment technologies to remove arsenic and mercury from contaminated groundwater: An overview. In Proceedings of the Southeast Asian Water Environment 1: Selected Papers from the First International Symposium on Southeast Aian Water Environment (biodiversity and Water Environment), Bangkok, Thailand, 1 December 2005.
14. Coupal, B.; Lalancette, J.-M. The treatment of waste waters with peat moss. *Water Res.* **1976**, *10*, 1071–1076. [CrossRef]
15. McLellan, J.; Rock, C. Pretreating landfill leachate with peat to remove metals. *Water Air Soil Pollut.* **1988**, *37*, 203–215. [CrossRef]
16. Chaney, R.L.; Hundemann, P.T. Use of peat moss columns to remove cadmium from wastewater. *J. Water Pollut. Control Fed.* **1979**, *51*, 17–21.
17. Sharma, D.; Forster, C. Removal of hexavalent chromium using sphagnum moss peat. *Water Res.* **1993**, *27*, 1201–1208. [CrossRef]
18. Gardea-Torresdey, J.; Tang, L.; Salvador, J. Copper adsorption by esterified and unesterified fractions of Sphagnum peat moss and its different humic substances. *J. Hazard. Mater.* **1996**, *48*, 191–206. [CrossRef]
19. Zhipei, Z.; Junlu, Y.; Zenghui, W.; Piya, C. A preliminary study of the removal of Pb(2+), Cd(2+), Zn(2+), Ni(2+), and Cr(2+) from wastewater with several Chinese peats. In Proceedings of the Seventh International Peat Congress, Dublin, Ireland, 18–23 June 1984; pp. 147–152.
20. Dennehy, C.; Lawlor, P.G.; Jiang, Y.; Gardiner, G.E.; Xie, S.; Nghiem, L.D.; Zhan, X. Green-house gas emissions from different pig manure management techniques: A critical analysis. *Front. Environ. Sci. Eng.* **2017**, *11*, 11.
21. Feng, Z.; Zhu, L. Sorption of phenanthrene to biochar modified by base. *Front. Environ. Sci. Eng.* **2017**, *12*, 1. [CrossRef]
22. Li, Z.; Wang, F.; Bai, T.; Tao, J.; Guo, J.; Yang, M.; Wang, S.; Hu, S. Lead immobilization by geological fluorapatite and fungus Aspergillus niger. *J. Hazard. Mater.* **2016**, *320*, 386–392. [CrossRef]
23. Shen, Z.; Zhang, Y.; Jin, F.; McMillan, O.; Al-Tabbaa, A. Qualitative and quantitative characterisation of adsorption mechanisms of lead on four biochars. *Sci. Total Environ.* **2017**, *609*, 1401–1410. [CrossRef]
24. Lee, S.-J.; Park, J.H.; Ahn, Y.-T.; Chung, J.W. Comparison of Heavy Metal Adsorption by Peat Moss and Peat Moss-Derived Biochar Produced Under Different Carbonization Conditions. *Water Air Soil Pollut.* **2015**, *226*, 9. [CrossRef]
25. Teir, S.; Eloneva, S.; Fogelholm, C.-J.; Zevenhoven, R. Stability of calcium carbonate and magnesium carbonate in rainwater and nitric acid solutions. *Energy Convers. Manag.* **2006**, *47*, 3059–3068. [CrossRef]
26. Petrounias, P.; Rogkala, A.; Giannakopoulou, P.P.; Tsikouras, B.; Lampropoulou, P.; Kalaitzidis, S.; Hatzipanagiotou, K.; Lambrakis, N.; Christopoulou, M.A. An Experimental Study for the Remediation of Industrial Waste Water Using a Combination of Low Cost Mineral Raw Materials. *Minerals* **2019**, *9*, 207. [CrossRef]
27. Smith, M. Panasqueira the tungsten giant at 100+. *Oper. Focus Int. Min.* **2006**, *33*, 10–14.

28. Brigatti, M.F.; Galán, E.; Theng, B. Chapter 2 Structures and Mineralogy of Clay Minerals. In *Developments in Clay Science*; Elsevier: Amsterdam, The Netherlands, 2006; Volume 1, pp. 19–86, ISBN 9780080441832.2006.
29. Sengupta, A.; Kadam, R.; Rajeswari, B.; Dhobale, A.; Babu, Y.; Godbole, S. Characterization of Indian serpentine by X-ray diffraction, photoacoustic spectroscopy and electron paramagnetic resonance spectroscopy. *Appl. Clay Sci.* **2010**, *50*, 305–310. [CrossRef]
30. Auzende, A.-L.; Pellenq, R.J.-M.; Devouard, B.; Baronnet, A.; Grauby, O. Atomistic calculations of structural and elastic properties of serpentine minerals: The case of lizardite. *Phys. Chem. Miner.* **2006**, *33*, 266–275. [CrossRef]
31. Huang, P.; Li, Z.; Chen, M.; Hu, H.; Lei, Z.; Zhang, Q.; Yuan, W. Mechanochemical activation of serpentine for recovering CU (II) from waste water. *Appl. Clay Sci.* **2017**, *149*, 1–7. [CrossRef]
32. Rogkala, A.; Petrounias, P.; Tsikouras, B.; Hatzipanagiotou, K. New Occurrence of Pyroxenites in the Veria-Naousa Ophiolite (North Greece): Implications on Their Origin and Petrogenetic Evolution. *Geoscience* **2017**, *7*, 92. [CrossRef]
33. Brunn, J.H. *Geological Map of Greece, Veroia Sheet, 1:50.000*; IGME: Athens, Greece, 1982.
34. Decourt, J.; Aubouin, J.; Savoyat, E. Le sillon mesohellenique et la zone pelagonienne. *Bull. Soc. Geoli. Fr.* **1977**, *1*, 32–70.
35. Michailidis, K. Zoned chromites with high Mn-contents in the Fe-Ni-Cr-laterite ore deposits from the Edessa area in Northern Greece. *Miner. Depos.* **1990**, *25*, 190–197. [CrossRef]
36. Rogkala, A.; Petrounias, P.; Tsikouras, B.; Giannakopoulou, P.P.; Hatzipanagiotou, K. Mineralogical Evidence for Partial Melting and Melt-Rock Interaction Processes in the Mantle Peridotites of Edessa Ophiolite (North Greece). *Minerals* **2019**, *9*, 120. [CrossRef]
37. Tarney JPe-Piper, G.; Piper, D.J.W. The Igneous Rocks of Greece. The Anatomy of an Orogen. Beiträge zur Regionalen Geologie der Erde (Series). *Geol. Mag.* **2003**, *140*, 357. [CrossRef]
38. Saccani, E.; Photiades, A.; Santato, A.; Zeda, O. New evidence for supra-subduction zone ophiolites in the Vardar zone of northern Greece: Implications for the tectonomagmatic evolution of the Vardar oceanic basin. *Ofioliti* **2008**, *33*, 65–85.
39. Mercier, J.L.; Vergely, P. *Geological Map of Greece, Edhessa Sheet, 1:50.000*; IGME: Athens, Greece, 1984.
40. Bish, D.L.; Post, J.E. Quantitative mineralogical analysis using the Rietveld full pattern fitting method. *Am. Mineral.* **1993**, *78*, 932–940.
41. Singh, A.K.; Singh, R.B. Genetic implications of Zn- and Mn-rich Cr-spinels in serpentinites of the Tidding Suture Zone, eastern Himalaya, NE India. *Geol. J.* **2012**, *48*, 22–38. [CrossRef]
42. Chen, M.; Li, Z.; Li, X.; Qu, J.; Zhang, Q. Mechanochemically extracting tungsten through caustic processing of scheelite by controlling calcium dissolution. *Int. J. Refract. Met. Hard Mater.* **2016**, *58*, 211–215. [CrossRef]
43. James, S.L.; Adams, C.J.; Bolm, C.; Braga, D.; Collier, P.; Friščić, T.; Grepioni, F.; Harris, K.D.M.; Hyett, G.; Jones, W.; et al. Mechanochemistry: Opportunities for new and cleaner synthesis. *Chem. Soc. Rev.* **2012**, *41*, 413–447. [CrossRef]
44. Vdović, N.; Jurina, I.; Škapin, S.D.; Sondi, I. The surface properties of clay minerals modified by intensive dry milling—Revisited. *Appl. Clay Sci.* **2010**, *48*, 575–580. [CrossRef]
45. Li, J.; Hitch, M. Mechanical activation of ultramafic mine waste rock in dry condition for enhanced mineral carbonation. *Miner. Eng.* **2016**, *95*, 1–4. [CrossRef]
46. Vijayaraghavan, K.; Yun, Y.-S. Bacterial biosorbents and biosorption. *Biotechnol. Adv.* **2008**, *26*, 266–291. [CrossRef]
47. Dutta, D.; Kumari, A.; Panda, R.; Jha, S.; Gupta, D.; Goel, S.; Jha, M.K. Close loop separation process for the recovery of Co, Cu, Mn, Fe and Li from spent lithium-ion batteries. *Sep. Purif. Technol.* **2018**, *200*, 327–334. [CrossRef]
48. Petrounias, P.; Giannakopoulou, P.P.; Rogkala, A.; Stamatis, P.M.; Tsikouras, B.; Papoulis, D.; Lampropoulou, P.; Hatzipanagiotou, K. The Influence of Alteration of Aggregates on the Quality of the Concrete: A Case Study from Serpentinites and Andesites from Central Macedonia (North Greece). *Geosciences* **2018**, *8*, 115. [CrossRef]
49. Petrounias, P.; Giannakopoulou, P.P.; Rogkala, A.; Stamatis, P.M.; Lampropoulou, P.; Tsikouras, B.; Hatzipanagiotou, K. The Effect of Petrographic Characteristics and Physico-Mechanical Properties of Aggregates on the Quality of Concrete. *Minerals* **2018**, *8*, 577. [CrossRef]

Article

NO_x Emissions and Nitrogen Fate at High Temperatures in Staged Combustion

Song Wu [1,*], Defu Che [2], Zhiguo Wang [1] and Xiaohui Su [1]

1 School of Mechanical Engineering, Xi'an Shiyou University, Xi'an 710065, China; zhgwang@xsyu.edu.cn (Z.W.); xhsu@xsyu.edu.cn (X.S.)

2 State Key Laboratory of Multiphase Flow in Power Engineering, School of Energy and Power Engineering, Xi'an Jiaotong University, Xi'an 710049, China; dfche@mail.xjtu.edu.cn

* Correspondence: songwu@xsyu.edu.cn

Received: 4 June 2020; Accepted: 8 July 2020; Published: 10 July 2020

Abstract: Staged combustion is an effective technology to control NO_x emissions for coal-fired boilers. In this paper, the characteristics of NO_x emissions under a high temperature and strong reducing atmosphere conditions in staged air and O_2/CO_2 combustion were investigated by CHEMKIN. A methane flame doped with ammonia and hydrogen cyanide in a tandem-type tube furnace was simulated to detect the effects of combustion temperature and stoichiometric ratio on NO_x emissions. Mechanism analysis was performed to identify the elementary steps for NO_x formation and reduction at high temperatures. The results indicate that in both air and O_2/CO_2 staged combustion, the conversion ratios of fuel-N to NO_x at the main combustion zone exit increase as the stoichiometric ratio rises, and they are slightly affected by the combustion temperature. The conversion ratios at the burnout zone exit decrease with the increasing stoichiometric ratio at low temperatures, and they are much higher than those at the main combustion zone exit. A lot of nitrogen compounds remain in the exhaust of the main combustion zone and are oxidized to NO_x after the injection of a secondary gas. Staged combustion can lower NO_x emissions remarkably, especially under a high temperature (≥1600 °C) and strong reducing atmosphere (SR ≤ 0.8) conditions. Increasing the combustion temperature under strong reducing atmosphere conditions can raise the H atom concentration and change the radical pool composition and size, which facilitate the reduction of NO to N_2. Ultimately, the increased OH/H ratio in staged O_2/CO_2 combustion offsets part of the reducibility, resulting in the final NO_x emissions being higher than those in air combustion under the same conditions.

Keywords: chemical simulation; NO_x emission; staged combustion; high temperature; strong reducing atmosphere

1. Introduction

Nitrogen oxides (NO_x) are one of the most predominant pollutants in coal-fired boilers. Because NO_x can endanger human health severely and cause acid rain, there is an increasing public demand for reducing NO_x emissions. Many countries have promulgated new NO_x emission limits [1–4]. In China, the allowed NO_x emissions should be below 100 mg of NO_2/Nm^3 (6% O_2) for all coal-fired power plants after 2014. In the European Union, the NO_x emission limit is expected to be lowered to 200 mg of NO_2/Nm^3 (6% O_2) for power plants over 500 MW_e by the year of 2016.

To achieve these stringent NO_x emission limits, a combination of two or more NO_x reduction techniques has to be used [5,6]. Currently, commercially available NO_x reduction techniques include air staging, reburning, low-nitrogen burner [7], selective catalytic reduction (SCR) and selective non-catalytic reduction (SNCR). Among them, air staging, reburning and low-nitrogen burner are low NO_x combustion techniques, whereas SCR and SNCR are post-combustion NO_x reduction technologies employed to offer varying degrees of NO_x control capability. Because of high capital and operating costs, ammonia leakage, deactivation of the catalyst, and relatively narrow temperature windows for post-combustion NO_x-reduction technologies [8–13], it is more desirable in practice to remove as much NO_x as possible during combustion to alleviate the dependency on SCR and SNCR.

Air staging is the most widely used NO_x control technique during combustion in pulverized coal boilers, which employs a main combustion zone and a burnout zone within the furnace region (see Figure 1) [5,14,15]. About 80% of air flow enters the main combustion zone where pulverized coal burns under substoichiometric conditions, i.e., reducing atmosphere conditions. The balance of the required air for complete combustion is introduced to the burnout zone through overfire air (OFA) ports. In the main combustion zone, the deficiency of oxygen can accelerate CH_m (with $m = 1, 2$, and 3) and NH_n (with $n = 0, 1$, and 2) to reduce generated NO_x to N_2 [16]. After the balance of oxygen is supplied to the burnout zone, a small amount of fuel NO_x will be formed here, as most of the fuel-N has been converted into N_2 or NO_x. Meanwhile, the generation of thermal NO_x is also inhibited. As a result, lower emissions of NO_x are achieved at the furnace exit.

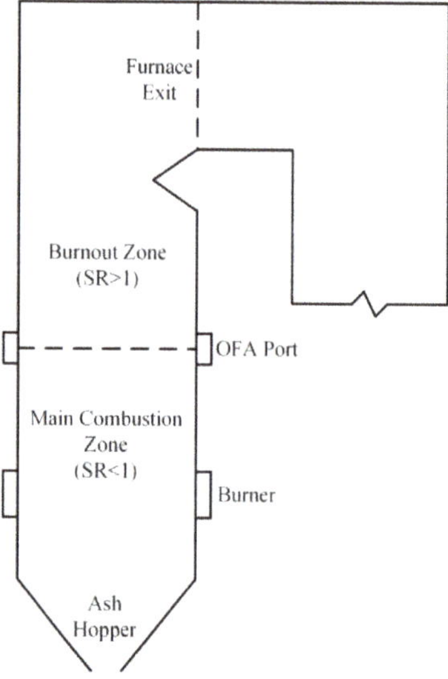

Figure 1. Pulverized coal boiler air-staged combustion NO_x control system.

Many research efforts have been devoted to the characteristics of NO_x formation and reduction in staged air combustion [1,3,16–22]. The results revealed that NO_x emissions are mostly determined by coal properties, combustion temperature, stoichiometric ratio (SR) and residence time. Some investigators [17,22] reported that stoichiometric ratio can strongly affect the generation of NO_x, and the conversion ratio of fuel-N to NO_x falls with the decreasing stoichiometric ratio under reducing atmosphere conditions (SR < 1). Taniguchi et al. [1,19] detected the impact of combustion temperature on NO_x emissions in a drop-tube furnace system used to simulate the air-staged combustion characteristics of actual pulverized coal boilers. They agreed that raising the temperature of the main combustion zone under reducing atmosphere conditions helps the reduction of NO_x. This viewpoint is also supported by some other research work [21]. Thus, it is found that a high temperature or strong reducing atmosphere in the main combustion zone can lower NO_x emissions. In other words, to achieve the minimal NO_x emissions in staged air combustion, the combustion temperature should be raised as high as possible and the stoichiometric ratio should be reduced as far as possible simultaneously in the main combustion zone. Bai et al. [16] examined NO_x emission levels of various coals under high temperature and strong reducing atmosphere conditions in a vertical tandem-type drop-tube furnace system. Their results verified the NO_x removal potential of this combustion method.

Creating high temperature and strong reducing atmosphere conditions in staged air combustion to lower NO_x emissions has a promising prospect. The slag-tap furnace is expected to be the most appropriate application for the combustion method owing to the following two aspects: The first is the high combustion temperature. For example, the gas temperature within the cyclone barrel is more than 1650 °C in cyclone-fired boilers [23]. The second is the decreased ash melting temperature under strong reducing atmosphere conditions [24]. The slag-tap furnace prefers to fire the coals having low ash melting temperatures without severe slagging. Moreover, there are also some other furnaces featuring high combustion temperatures [5,21].

O_2/CO_2 combustion is one of the most promising CO_2-capture technologies in thermal power generation [25–28]. Studies have shown that staged combustion can also lower the NO_x emissions in O_2/CO_2 combustion [16,20,26,29]. Thus, NO_x formation and reduction under high temperatures and strong reducing atmosphere conditions in staged O_2/CO_2 combustion are also worthy being explored.

A number of researchers [4,6,20,30–35] have investigated the processes of NO_x formation and reduction in staged air and O_2/CO_2 combustion to date. However, these studies were performed under the conditions of relatively low temperature or relatively high stoichiometric ratio. There are very few studies on the reaction mechanisms for NO_x under high temperatures and strong reducing atmosphere conditions, which differ from those under conventional conditions of staged combustion. Meanwhile, how to co-ordinate the combustion temperature and stoichiometric ratio to achieve the most suitable conditions for NO_x reduction is still unclear and insufficiently understood. Therefore, it is of great significance to study how the high temperature and strong reducing atmosphere conditions influence the NO_x formation and reduction in staged combustion.

In the present study, the characteristics of NO_x formation and reduction under high temperature and strong reducing atmosphere conditions in staged air combustion were investigated numerically by CHEMKIN. A methane flame doped with ammonia and hydrogen cyanide for fuel-N in a tandem-type tube furnace was simulated to probe the effects of combustion temperature and stoichiometric ratio on the NO_x emissions. Based on the calculations, the elementary steps for NO_x formation and reduction at the high temperature were identified. In addition, NO_x formation and reduction in staged O_2/CO_2 combustion were also examined.

2. Numerical Approach

2.1. Reactors and Models

The simulations were carried out for a tandem-type tube furnace consisting of two identical tube reactors. The tube furnace is shown schematically in Figure 2a. The inside diameter and heating length of the tube reactor are 38 mm and 600 mm, respectively. The heaters are arranged around the tube reactors to control the reaction temperatures. The primary gas, which includes CH_4, NH_3, HCN and air (O_2/CO_2), is introduced into the tube reactor 1 where methane burns under different temperatures and stoichiometric conditions. Meanwhile, NH_3 and HCN are converted into NO or N_2. Subsequently, the secondary gas, i.e., supplementary air or O_2/CO_2, is injected as OFA from the connection between the tube reactor 1 and 2. The residual fuel burns out in the tube reactor 2. Therefore, the tube furnace can be employed to describe the characteristics of NO_x formation and reduction in staged combustion with the tube reactor 1 and 2 regarded as the main combustion zone and the burnout zone, respectively.

Indeed, some key processes in real boilers such as strong turbulence, devolatilization, char combustion and thermal radiation are simplified for the purpose of seeking elementary steps for N conversion at high temperatures. During pulverized coal combustion, considerable fractions of C and N conversion occur in the gas phase. When pulverized coal is ignited in the furnace, the volatiles in coal are first released and then mixed with air for homogeneous combustion. Hydrocarbons are the important components of the volatiles. Compared with the real combustion process of pulverized coal, the transformation of Char-N and heterogeneous reduction of NO_x are not considered. This simplification certainly brings deviation for direct prediction of NO_x emissions in practice, and its influence depends on the amount of the coal volatiles. In this study, although methane combustion simulation cannot fully reflect the N conversion during coal combustion, it can still reveal NO_x formation and reduction in the homogeneous combustion.

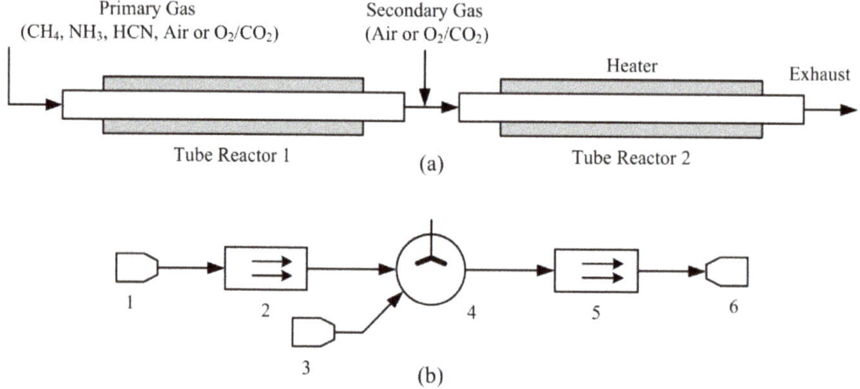

Figure 2. Schematic diagram of simulation object: (**a**) tube furnace of staged combustion; (**b**) reaction process modeling, 1—external source of inlet gas i, 2—plug flow reactor i, 3—external source of inlet gas ii, 4—non-reactive gas mixer, 5—plug flow reactor ii, 6—outlet flow of reactor.

The simulations were performed using a chemical kinetics modeling code CHEMKIN. It provides a feasible and powerful tool to understand reaction processes involving elementary gas-phase chemical kinetics [30–32,34,36–38]. Proper models were chosen to simulate the tube furnace and a corresponding reaction process diagram was developed in Figure 2b. Two external source of inlet gas models were used to introduce the primary and secondary gases into the reaction system. Moreover, two plug flow reactor (PFR) models were employed to describe the combustion processes in the tube reactor 1 and 2, respectively. The PFR model assumes that no mixing occurs in the flow direction while perfect mixing occurs in the direction perpendicular to the flow [39,40]. Many researchers have applied it to simulate the complex physical and chemical phenomena in tube reactors. A non-reactive gas mixer model was used to replace the connection between the two tube reactors. The reaction system ended up with an outlet flow of reactor model. All the models above completely constituted the simulation object.

2.2. Reaction Mechanism

There are three mechanisms responsible for the NO_x formation in combustion systems: thermal NO_x, prompt NO_x and fuel NO_x. In the present study, the production of NO is far more than those of other nitrogen oxides, thus, only NO is taken into account in our results. The thermal NO_x is formed by the direct oxidation of nitrogen from the combustion air at a very high temperature (>1800 K). This reaction process can be expressed by the extended Zeldovich mechanism as follows [41]:

$$N_2 + O \rightarrow NO + N \tag{1}$$

$$N + O_2 \rightarrow NO + O \tag{2}$$

$$N + OH \rightarrow NO + H \tag{3}$$

The prompt NO_x are generated by the reaction of atmospheric nitrogen with hydrocarbon radicals in fuel-rich conditions. The key reactions are written by [42]:

$$N_2 + CH \rightarrow HCN + N \tag{4}$$

$$N_2 + CH_2 \rightarrow HCN + NH \tag{5}$$

Subsequently, these resultants are oxidized to NO quickly. The fuel NO_x are produced by the oxidation of nitrogen bound in the fuel and generally account for more than 80% of the total NO_x production in large pulverized coal boilers [41]. NH_3 and HCN are the dominant intermediates during the conversion of fuel-N to NO or N_2. As a result, the overall reactions of the fuel NO_x formation can be given by:

$$NH_3 + O_2 \rightarrow NO + \ldots \tag{6}$$

$$HCN + O_2 \rightarrow NO + \ldots \tag{7}$$

In addition, the generated NO is also reduced to N_2 simultaneously, which mainly depends on the local environment. The final NO_x emissions are the comprehensive result of these actions, and a detailed reaction mechanism is needed to predict it.

A GRI-Mech 3.0 reaction mechanism was adopted in this paper, which involves 53 species and 325 elementary chemical reactions [43]. The purpose of this mechanism is to model natural gas combustion, including NO formation and reduction and reburn chemistry. The three NO_x formation mechanisms above are all included in this mechanism. Species concentrations in reaction systems are calculated from the net rate of production for each species by chemical reaction. Reaction rate constants are determined by the modified Arrhenius expression [32]:

$$k = AT^{\beta} \exp(-E/[RT]) \tag{8}$$

where A is the pre-exponential factor, T is the reaction temperature, β is the correction factor, E is the activation energy and R is the molar gas constant. The reverse reaction rate constants derive from the forward reaction rate constants and appropriate equilibrium constants. Under this mechanism, the rate of production (ROP) and first-order sensitivity analyses were used to interpret the kinetic results [32,34,36,44]. The ROP analysis can provide the information of the rates of formation and consumption for each species involved in the mechanism. The first-order sensitivity analysis is able to obtain the first-order sensitivity coefficient defined as:

$$\kappa = \frac{\delta Y_j / Y_j}{\delta A_i / A_i} \tag{9}$$

where Y_j is the mole fraction for the jth species and A_i is the pre-exponential factor for the ith reaction. The coefficient reflects the relative change in the predicted concentration for jth species caused by increasing the reaction rate constant for the ith reaction.

3. Data Analysis and Simulation Conditions

The stoichiometric ratio (SR) is often used to express combustion conditions, which is represented in this study by:

$$SR = \frac{V_{O_2}/V_{CH_4}}{\left(V_{O_2}/V_{CH_4}\right)_{st}} \tag{10}$$

where V is the volume flow rate and the subscript st denotes the stoichiometric condition. The conversion ratio of fuel-N to NO_x (NO_x CR) is defined as:

$$NO_x\ CR = \frac{\text{Exhaust } NO_x \text{ volume flow rate}}{\text{Inflow fuel} - \text{N volume flow rate}} \tag{11}$$

where the Inflow fuel-N volume flow rate is the sum of NH_3 and HCN volume flow rates.

Nine different stoichiometric ratios in the main combustion zone (0.5–1.2, 2) were used to study the effect of the SR on the characteristics of NO_x emissions, as shown in Table 1. The C/N mole ratios in these cases were all selected as 85. The reaction temperature in the main combustion zone varies from 1200 to 1800 °C, while that in the burnout zone varies from 1100 to 1400 °C. To compare with air-staged combustion, NO_x emissions in staged O_2/CO_2 and O_2/Ar combustion were also investigated with the O_2 concentrations set at the same value (21%).

Table 1. Simulation conditions for staged combustion in the tandem-type tube furnace.

	SR (Main Combustion Zone)								
	0.5	0.6	0.7	0.8	0.9	1	1.1	1.2	2
Primary gas (Ncm^3/min):									
CH_4	735	612.5	525	459.4	408.3	367.5	334.1	306.2	183.8
O_2	735	735	735	735	735	735	735	735	735
N_2 (CO_2 or Ar)	2765	2765	2765	2765	2765	2765	2765	2765	2765
NH_3	3.46	2.88	2.47	2.16	1.92	1.73	1.57	1.44	0.86
HCN	5.19	4.32	3.71	3.24	2.88	2.59	2.36	2.16	1.30
Secondary gas (Ncm^3/min):									
O_2	1029	735	525	367.5	245	147	66.8	0	0
N_2 (CO_2 or Ar)	3871	2765	1975	1382.5	921.7	553	251.4	0	0
SR_{global}	1.2	1.2	1.2	1.2	1.2	1.2	1.2	1.2	2
Reaction temperature: 1200–1800 °C (Main combustion zone); 1100–1400 °C (Burnout zone)									

4. Results and Discussion

4.1. Model Validation

In order to obtain creditable and reasonable simulation results, a comparison between different reaction mechanisms and models was carried out in Figure 3. An updated reaction mechanism of Glarborg et al. and a premixed flame model (PFM) were also taken into consideration in the present study. The updated Glarborg reaction mechanism includes 97 species and 779 elementary chemical reactions, and is able to predict the experimental results correctly [32]. The PFM can compute species and temperature profiles in steady-state burner-stabilized premixed laminar flames. Figure 3 gives the predicted results of three cases: PFR model and GRI-Mech 3.0 reaction mechanism, PFR model and updated Glarborg reaction mechanism, PFM model and GRI-Mech 3.0 reaction mechanism [45]. The calculation was conducted under the SR of 0.7 and the reaction temperature varying from 1200 to 1800 °C; the other conditions are listed in Table 1. On the whole, the predicted volume flow rates of NH_3, HCN and NO at the main combustion zone exit show similar trends in the three cases. Although there are some differences between the reaction mechanisms and models, these results are comparable. Especially at high temperatures, a good agreement is observed. Therefore, the above-described numerical approach is valid.

4.2. Characteristics of NO_x Emissions in Staged Air Combustion

Figure 4 shows the NO_x CRs at the main combustion zone and burnout zone exits in the staged air combustion. The NO_x CRs at the main combustion zone exit increase when the SR rises, and the reaction temperature seems to have little influence on the NO_x CRs. However, the NO_x CRs at the burnout zone exit are significantly affected by the SR and reaction temperature in the main combustion zone. When the reaction temperature is low, the NO_x CRs decrease with the increasing SR. Comparing the NO_x CRs of the main combustion zone exit and the burnout zone exit, a large quantity of NO_x is produced after the injection of secondary gas, which means a lot of nitrogen compounds exist in the exhaust of the main combustion zone. This point will be proved and discussed in Figure 5. The smaller the SR is, the higher the number of nitrogen compounds. Contrastingly, for the high reaction temperature, the final NO_x emission levels are quite low. A minimal difference is found in the NO_x CRs between the main combustion zone exit and the burnout zone exit, which means most of the fuel-N has been converted into NO_x or N_2 and few nitrogen compounds remain at the main combustion zone exit. In addition, the lower NO_x emission levels are found at a high temperature (≥1600 °C) and strong reducing atmosphere (SR ≤ 0.8) conditions.

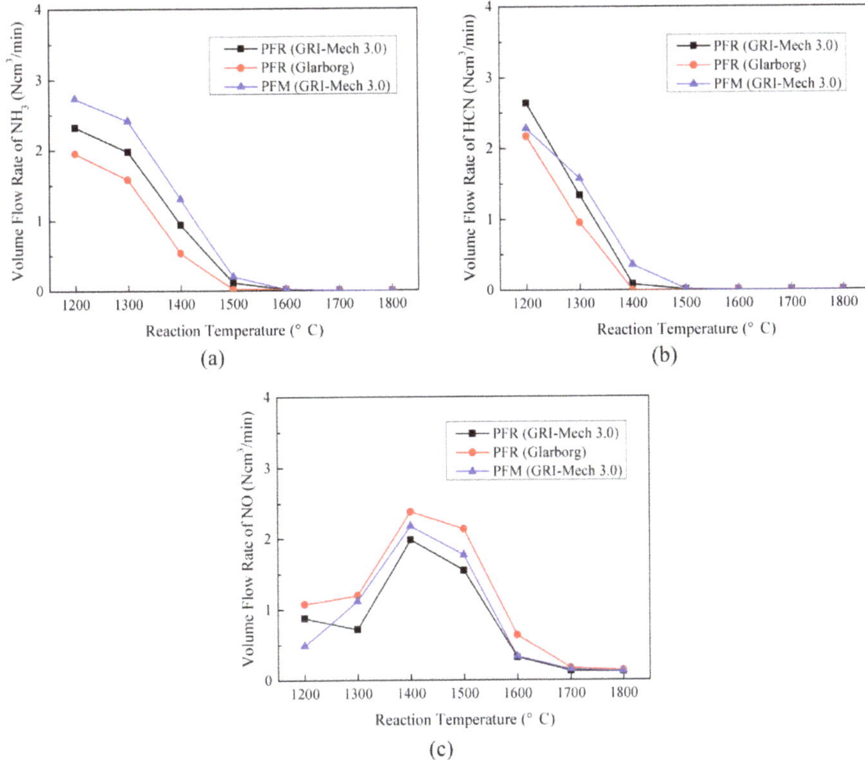

Figure 3. Comparison of different reaction mechanisms and models in staged air combustion: (**a**–**c**) are the predicted volume flow rates of NH_3, HCN and NO, respectively, at the main combustion zone exit under the SR of 0.7.

The sums of NH_3, HCN, and NO_x at the main combustion zone and burnout zone exits in staged air combustion are shown in Figure 5. At the main combustion zone exit, the NO_x emissions are low, while the sum of NH_3, HCN, and NO_x is quite high when the reaction temperature is set as 1200 °C. A large amount of NH_3 and HCN remain in the primary combustion exhaust. With the secondary gas introduced, the remaining NH_3 and HCN are almost entirely oxidized to NO_x. Therefore, at the burnout zone exit, very little NH_3 and HCN exist, and the sum of NH_3, HCN, and NO_x is approximately equal to the NO_x emissions. The final NO_x emissions depend on the sum of nitrogen compounds in the primary combustion exhaust. To limit the NO_x emissions as much as possible by air staging, it is of great significance to obtain a minimum sum of NH_3, HCN, and NO_x in the main combustion zone, i.e., converting more fuel-N to N_2 in terms of N balance. With the reaction temperature in the main combustion zone rising, the final NO_x emissions decrease. When the temperature is higher than 1600 °C, the final NO_x emissions attain a minimum level.

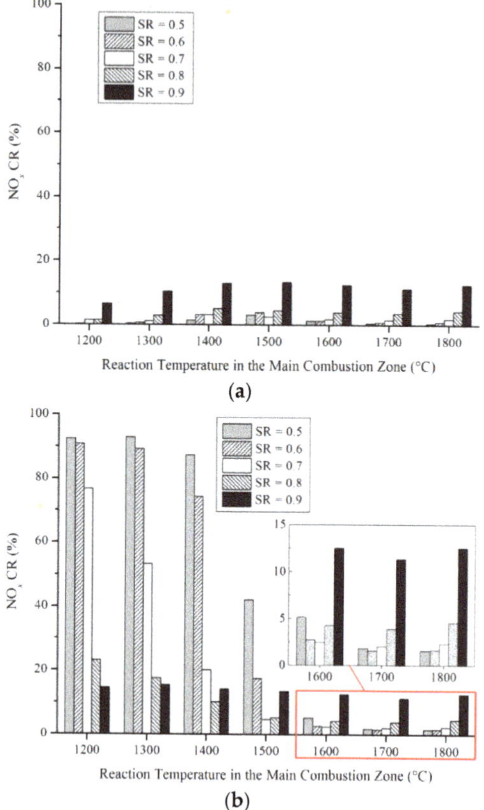

Figure 4. Conversion ratio of fuel-N to NO_x in staged air combustion (reaction temperature in the burnout zone: 1100 °C): (**a**) main combustion zone exit; (**b**) burnout zone exit.

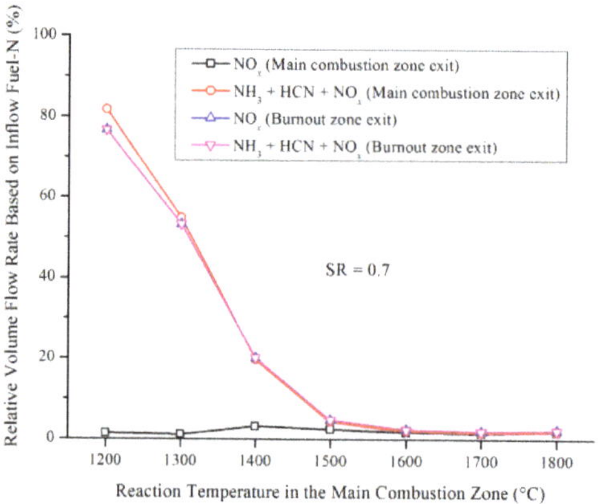

Figure 5. Sums of NH_3, HCN and NO_x at the exits in staged air combustion (reaction temperature in the burnout zone: 1100 °C).

Figure 6 illustrates the NO_x emissions under the oxidizing atmosphere condition (SR ≥ 1) in air combustion. Considering the formation of the thermal NO_x, a simulation of O_2/Ar combustion was performed for comparison. Most of the fuel-N is easily oxidized to NO_x by the excess O_2. When the combustion temperature is higher than 1500 °C, the thermal NO_x begins to be markedly produced. As the combustion temperature and SR rise, the emissions of the thermal NO_x increase rapidly.

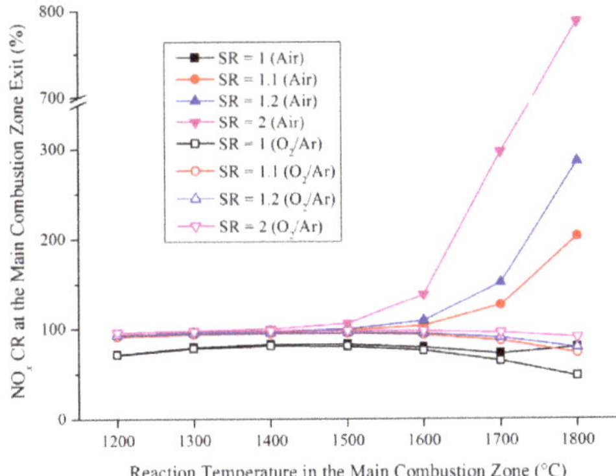

Figure 6. Comparison of NO_x emissions between air combustion and O_2/Ar combustion under oxidizing atmosphere conditions.

The effect of the reaction temperature in the burnout zone (T_2) on the NO_x emissions in air staging is shown in Figure 7. The reaction temperature in the main combustion zone (T_1) is selected as 1500 °C. There is minimal difference in the final NO_x CR when T_2 varies from 1100 to 1400 °C. In other words, T_2 nearly has no effect on the formation of NO_x. Therefore, it is important to control the NO_x formation and reduction in the main combustion zone instead of the burnout zone.

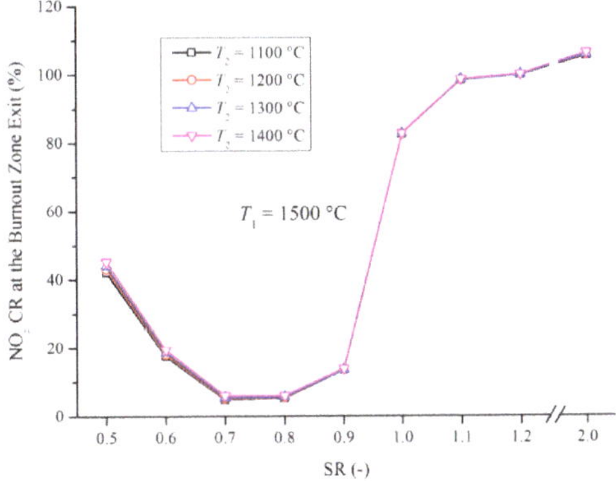

Figure 7. Effect of reaction temperature in the burnout zone on the NO_x emissions in staged air combustion.

4.3. Characteristics of NO_x Emissions in Staged O_2/CO_2 Combustion

Figure 8 presents the effects of the SR and combustion temperature on the NO_x emissions in the staged O_2/CO_2 combustion. The variation trends of the NO_x CRs are similar to those in the staged air combustion. However, a significant difference in the exact NO_x emission value between O_2/CO_2 combustion and air combustion is found due to the existence of a great deal of CO_2. In the O_2/CO_2 combustion, the CO_2 concentration is so high that the chemical reaction 12 is observably facilitated [30,37]:

$$H + CO_2 \rightarrow OH + CO \tag{12}$$

Here, CO_2 cannot be considered as an inert gas anymore. The chemical reaction 12 diminishes the H atom concentration and increases the concentration of OH radicals, which impacts the NO_x formation and reduction strongly. Compared with the staged air combustion, the NO_x CR at the main combustion zone exit increases markedly, and the range of the SR and combustion temperature (under which a significant amount of NH_3 and HCN remain in the primary combustion exhaust) is narrow. At low temperatures (≤1400 °C), a significant amount of NH_3 and HCN remain at the main combustion zone exit. Their sum rises rapidly with the decreasing SR and combustion temperature. Similarly, the lower final NO_x emission levels appear at high temperatures (≥1500 °C), and the higher combustion temperature and the smaller SR lead to lower NO_x emissions. Furthermore, the final NO_x emissions in the staged O_2/CO_2 combustion are higher than those in the staged air combustion at the same high temperature and strong reducing atmosphere conditions. This conclusion is consistent with Mendiara et al.'s research results [32,36]. Because of the chemical reaction 12, the OH/H ratio increases, which is equivalent to providing an oxidizing agent in the combustion atmosphere. Therefore, the oxidation of NH_3 and HCN to NO_x is promoted.

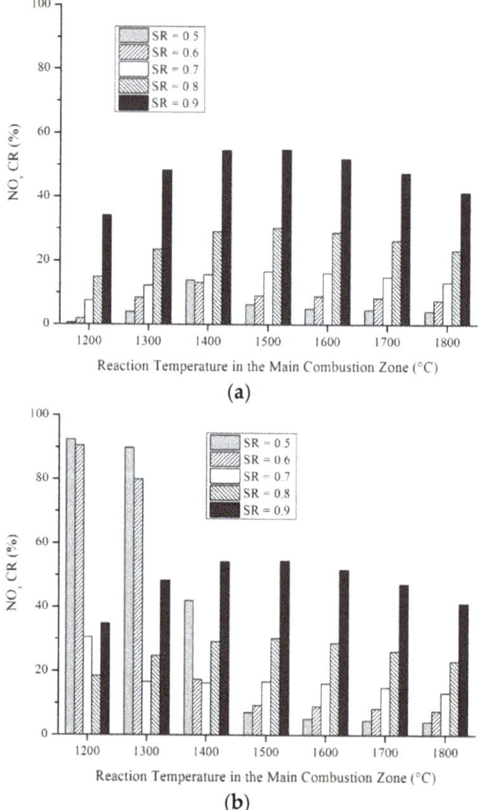

Figure 8. Conversion ratio of fuel-N to NO_x in staged O_2/CO_2 combustion (reaction temperature in the burnout zone: 1100 °C): (**a**) main combustion zone exit; (**b**) burnout zone exit.

Figure 9 compares the NO_x emissions of staged air and O_2/CO_2 combustion under different atmosphere conditions. In O_2/CO_2 combustion, staged combustion is also able to decrease the NO_x emissions enormously, but the emission reduction is less than that in air staging. Under the oxidizing atmosphere condition, the NO_x emission levels are quite high in both air combustion and O_2/CO_2

combustion. Moreover, CO_2 can reduce the O/H radical pool and tends to inhibit the NO_x formation from fuel-N and, thus, the NO_x emissions in O_2/CO_2 combustion are lower than those in air combustion. When under the reducing atmosphere condition, as the SR falls, the NO_x CR in O_2/CO_2 combustion decreases while that in air combustion reduces first and then increases. The NO_x CR of O_2/CO_2 combustion has a minimum of 7.1% at the SR of 0.5, while that of air combustion reaches a minimum of 4.7% at the SR of 0.7. These results denote that staged air combustion is more practical for limiting the NO_x emissions than staged O_2/CO_2 combustion.

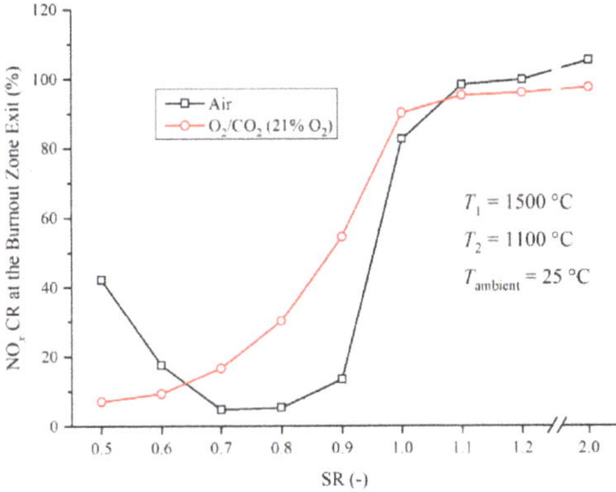

Figure 9. Comparison of NO_x emissions between air combustion and O_2/CO_2 combustion under different atmosphere conditions.

Another interesting finding in Figure 9 is that there are two reverse trends for staged air combustion and O_2/CO_2 combustion at a smaller SR. The chemical reaction 12 can affect reducibility of combustion atmosphere strongly. At a smaller SR, a certain amount of NH_3 and HCN remain in the primary combustion exhaust for staged air combustion when the temperature is not so high (here 1500 °C), then the NO_x CR increases after these NH_3 and HCN are oxidized to NO_x by the OFA. While for staged O_2/CO_2 combustion, the atmosphere is much less reductive at the same SR; only a small amount of NH_3 and HCN remain when the temperature is 1500 °C. Moreover, the smaller SR is, the more significant the effect of reducibility.

4.4. Mechanism Analysis

According to the ROP analysis, a reaction path diagram reflecting the main reaction pathways for the conversion of NH_3 and HCN to NO or N_2 in staged air and O_2/CO_2 combustion is proposed in Figure 10. The combustion temperature is 1600 °C and the SR is 0.7 during the calculation. The solid lines represent reaction pathways important in air combustion, while the dashed lines express those only significant in O_2/CO_2 combustion. It can be seen from the reaction path diagram that NO is directly reduced to N_2 mainly through the following reactions:

$$N + NO \rightarrow N_2 + O \tag{13}$$

$$NH + NO \rightarrow N_2 + OH \tag{14}$$

$$NCO + NO \rightarrow N_2 + CO_2 \tag{15}$$

Besides, part of NO first forms nitrogen intermediates NNH and N_2O, and they are then converted into N_2 by:

$$NH + NO \rightarrow NNH + O \tag{16}$$

$$NNH + H \rightarrow H_2 + N_2 \quad (17)$$

$$NNH + OH \rightarrow H_2O + N_2 \quad (18)$$

$$NNH \rightarrow N_2 + H \quad (19)$$

$$NH + NO \rightarrow N_2O + H \quad (20)$$

$$NCO + NO \rightarrow N_2O + CO \quad (21)$$

$$N_2O + H \rightarrow N_2 + OH \quad (22)$$

$$N_2O(+M) \rightarrow N_2 + O(+M) \quad (23)$$

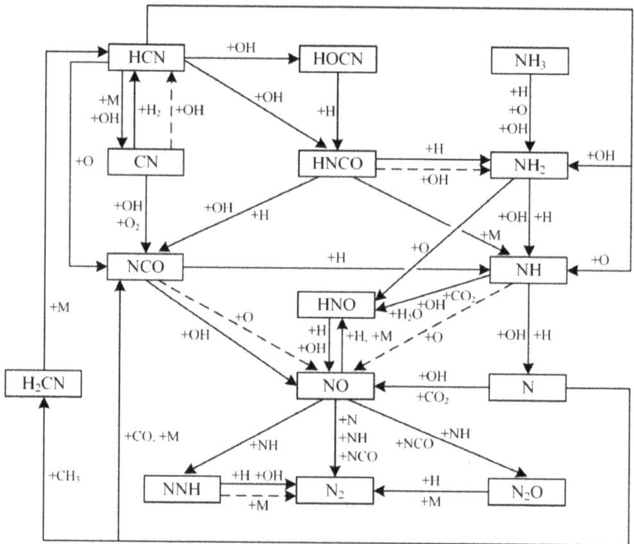

Figure 10. Reaction path diagram for the fuel-N conversion under a high temperature and strong reducing atmosphere conditions in staged air and O_2/CO_2 combustion.

Major reactions for NH_3 consumption are the interactions with H, O and OH radicals:

$$NH_3 + H \rightarrow NH_2 + H_2 \quad (24)$$

$$NH_3 + O \rightarrow NH_2 + OH \quad (25)$$

$$NH_3 + OH \rightarrow NH_2 + H_2O \quad (26)$$

Main reactions for HCN removal are listed as follows:

$$HCN + OH \rightarrow NH_2 + CO \quad (27)$$

$$HCN + O \rightarrow NH + CO \quad (28)$$

$$HCN + O \rightarrow NCO + H \quad (29)$$

$$HCN + OH \rightarrow HOCN + H \quad (30)$$

$$HCN + OH \rightarrow HNCO + H \quad (31)$$

$$HCN + M \rightarrow H + CN + M \quad (32)$$

$$HCN + OH \rightarrow CN + H_2O \quad (33)$$

There are also some important nitrogen intermediates formed during the conversion of NH_3 and HCN, such as NCO, NH, NH_2, and HNO. Whether the fuel-N is finally converted into NO or N_2 depends on the formation and evolution of these nitrogen intermediates, which are significantly affected by the presence of H, O and OH radicals in the reaction atmosphere. Determined by the combustion temperature and SR, different concentrations of H, O and OH radicals lead to different NO_x emissions. Under the high temperature and strong reducing atmosphere conditions, the H atom concentration is increased, and the OH/H ratio and O/H ratio are decreased correspondingly. As a result, the reactions by which the nitrogen intermediates are oxidized to NO are inhibited, while those promoting the NO reduction to N_2 are enhanced. For O_2/CO_2 combustion, reaction 12 is considered to be responsible for the impact of high CO_2 concentration. It can compete with reaction 34 for H [30,37], which changes the concentrations of H, O and OH radicals in the reaction atmosphere.

$$H + O_2 \rightarrow O + OH \quad (34)$$

Figure 11 gives the results of a first-order sensitivity analysis for N_2 at the SR of 0.7 in air combustion. Here, the effect of temperature on N_2 production is detected emphatically. The first-order sensitivity coefficients of some reactions are negative at a low temperature, while they become positive at a high temperature, which means that these reactions play important roles in the reduction of NO_x to N_2. With the increasing temperature, the function of these reactions switches from inhibiting N_2 production to facilitating it. This is because the increasing rate constant of each reaction induced by the higher temperature changes the radical pool composition and size in the reaction atmosphere. Moreover, N_2 is mostly sensitive to the reactions that generate or consume H and CH_3 radicals under the high temperature and strong reducing atmosphere conditions. For instance, increasing the rate of reaction 34 will promote NO_x reduction. Similarly, Figure 12 displays the first-order sensitivity analysis for N_2 in O_2/CO_2 combustion. Due to reaction 12, some reactions for CH_3 consumption become bottlenecks in N_2 formation, besides reaction 34.

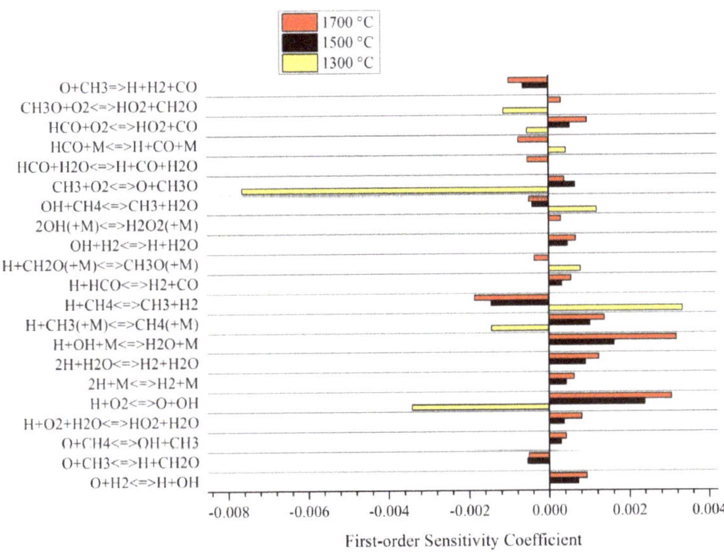

Figure 11. First-order sensitivity analysis for N_2 at different temperatures in air combustion (SR = 0.7).

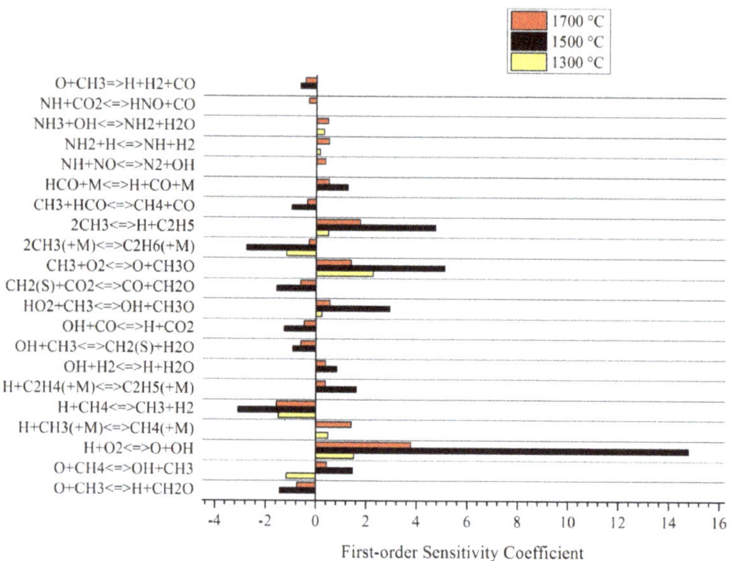

Figure 12. First-order sensitivity analysis for N_2 at different temperatures in O_2/CO_2 combustion (SR = 0.7).

5. Conclusions

In this study, a methane flame doped with ammonia and hydrogen cyanide for fuel-N in a tandem-type tube furnace was simulated to investigate the characteristics of NO_x emissions under a high temperature and strong reducing atmosphere conditions in staged air and O_2/CO_2 combustion by CHEMKIN. The effects of combustion temperature and stoichiometric ratio on the NO_x emissions were examined, and the elementary steps for NO_x formation and reduction at high temperatures were identified. The following conclusions can be drawn:

In both staged air and staged O_2/CO_2 combustion (SR < 1), the NO_x CRs at the main combustion zone exit increase as the SR rises, and they are slightly affected by the combustion temperature. The NO_x CRs at the burnout zone exit decrease with the increasing SR at low temperatures, and they are much higher than those at the main combustion zone exit. Here, a lot of nitrogen compounds remain in the exhaust of the main combustion zone and can be easily oxidized to NO_x with the injection of secondary gas. Staged combustion can lower the NO_x emission levels significantly, especially under a high temperature (≥1600 °C) and strong reducing atmosphere (SR ≤ 0.8) conditions. Increasing the combustion temperature under strong reducing atmosphere conditions can raise the H atom concentration and change the radical pool composition and size. Therefore, the reactions by which NO is reduced to N_2 are facilitated. In addition, the increased OH/H ratio through reaction 12 offsets part of the reducibility in staged O_2/CO_2 combustion, resulting in the final NO_x emissions in O_2/CO_2 combustion being higher than those in air combustion at the same high temperature and strong reducing atmosphere conditions.

Author Contributions: Conceptualization, S.W. and D.C.; methodology, S.W.; software, S.W. and Z.W.; validation, S.W. and X.S.; data curation, S.W. and D.C.; writing—original draft preparation, S.W. and D.C.; writing—review and editing, Z.W. and X.S. All authors have read and agreed to the published version of the manuscript.

Funding: This work has been financially supported by the National Natural Science Foundation of China (Grant No. 51906202) and the Basic Research Program of Natural Science of Shaanxi Province (No. 2019JQ-809).

Conflicts of Interest: The authors declare no conflict of interest.

Nomenclature

A	pre-exponential factor (L/(mol·s))
E	activation energy (kJ/mol)
k	reaction rate constant (L/(mol·s))
NO_x CR	conversion ratio of fuel-N to NO_x (%)
R	molar gas constant (kJ/(mol·K))
SR	stoichiometric ratio (-)
T	reaction temperature (K)
V	volume flow rate (Ncm^3/min)
Y	mole fraction (-)
β	correction factor (-)
κ	first-order sensitivity coefficient (-)
i	index for reaction
j	index for species
st	stoichiometric condition

References

1. Taniguchi, M.; Kamikawa, Y.; Tatsumi, T.; Yamamoto, K. Staged combustion properties for pulverized coals at high temperature. *Combust. Flame* **2011**, *158*, 2261–2271. [CrossRef]
2. Fan, W.; Li, Y.; Lin, Z.; Zhang, M. PDA research on a novel pulverized coal combustion technology for a large utility boiler. *Energy* **2010**, *35*, 2141–2148. [CrossRef]
3. Fan, W.; Lin, Z.; Li, Y.; Kuang, J.; Zhang, M. Effect of Air-Staging on Anthracite Combustion and NOxFormation. *Energy Fuels* **2009**, *23*, 111–120. [CrossRef]
4. Pisupati, S.V.; Bhalla, S. Numerical modeling of NOx reduction using pyrolysis products from biomass-based materials. *Biomass Bioenergy* **2008**, *32*, 146–154. [CrossRef]
5. Liu, H.; Liu, Y.; Yi, G.; Nie, L.; Che, D. Effects of Air Staging Conditions on the Combustion and NOx Emission Characteristics in a 600 MW Wall Fired Utility Boiler Using Lean Coal. *Energy Fuels* **2013**, *27*, 5831–5840. [CrossRef]
6. Molina, A.; Murphy, J.; Winter, F.; Haynes, B.; Blevins, L.; Shaddix, C. Pathways for conversion of char nitrogen to nitric oxide during pulverized coal combustion. *Combust. Flame* **2009**, *156*, 574–587. [CrossRef]
7. Zhang, Y.; Luo, R.; Dou, Y.; Zhou, Q. Combustion Characteristics and NOx Emission through a Swirling Burner with Adjustable Flaring Angle. *Energies* **2018**, *11*, 2173. [CrossRef]
8. Zhu, L.; Zhong, Z.; Yang, H.; Wang, C. Effect of MoO3 on vanadium based catalysts for the selective catalytic reduction of NOx with NH3 at low temperature. *J. Environ. Sci.* **2017**, *56*, 169–179. [CrossRef]
9. Yu, Y.; He, C.; Chen, J.; Yin, L.; Qiu, T.; Meng, X. Regeneration of deactivated commercial SCR catalyst by alkali washing. *Catal. Commun.* **2013**, *39*, 78–81. [CrossRef]
10. Kang, M.; Park, E.D.; Kim, J.M.; Yie, J.E. Manganese oxide catalysts for NOx reduction with NH3 at low temperatures. *Appl. Catal. A Gen.* **2007**, *327*, 261–269. [CrossRef]
11. Javed, M.T.; Irfan, N.; Gibbs, B. Control of combustion-generated nitrogen oxides by selective non-catalytic reduction. *J. Environ. Manag.* **2007**, *83*, 251–289. [CrossRef] [PubMed]
12. Bae, S.W.; Roh, S.; Kim, S.D. NO removal by reducing agents and additives in the selective non-catalytic reduction (SNCR) process. *Chemosphere* **2006**, *65*, 170–175. [CrossRef] [PubMed]
13. Forzatti, P. Present status and perspectives in de-NOx SCR catalysis. *Appl. Catal. A Gen.* **2001**, *222*, 221–236. [CrossRef]
14. Choi, C.R.; Kim, C.N. Numerical investigation on the flow, combustion and NOx emission characteristics in a 500MWe tangentially fired pulverized-coal boiler. *Fuel* **2009**, *88*, 1720–1731. [CrossRef]
15. Costa, M.; Azevedo, J. Experimental characterization of an industrial pulverized coal-fired furnace under deep staging conditions. *Combust. Sci. Technol.* **2007**, *179*, 1923–1935. [CrossRef]
16. Bai, W.; Li, H.; Deng, L.; Liu, H.; Che, D. Air-Staged Combustion Characteristics of Pulverized Coal under High Temperature and Strong Reducing Atmosphere Conditions. *Energy Fuels* **2014**, *28*, 1820–1828. [CrossRef]

17. Song, M.; Zeng, L.; Li, X.; Chen, Z.; Li, Z. Effect of Stoichiometric Ratio of Fuel-Rich Flow on Combustion Characteristics in a Down-Fired Boiler. *J. Energy Eng.* **2017**, *143*, 04016058. [CrossRef]
18. Yang, J.; Sun, R.; Sun, S.; Zhao, N.; Hao, N.; Chen, H.; Wang, Y.; Guo, H.; Meng, J. Experimental study on NOx reduction from staging combustion of high volatile pulverized coals. Part 1. Air staging. *Fuel Process. Technol.* **2014**, *126*, 266–275. [CrossRef]
19. Taniguchi, M.; Kamikawa, Y.; Tatsumi, T.; Yamamoto, K.; Kondo, Y. Relationships between Gas-Phase Stoichiometric Ratios and Intermediate Species in High-Temperature Pulverized Coal Flames for Air and Oxy-Fuel Combustions. *Energy Fuels* **2012**, *26*, 4712–4720. [CrossRef]
20. Taniguchi, M.; Kamikawa, Y.; Okazaki, T.; Yamamoto, K.; Orita, H. A role of hydrocarbon reaction for NOx formation and reduction in fuel-rich pulverized coal combustion. *Combust. Flame* **2010**, *157*, 1456–1466. [CrossRef]
21. Fan, W.; Lin, Z.; Li, Y.; Li, Y. Effect of Temperature on NO Release during the Combustion of Coals with Different Ranks. *Energy Fuels* **2010**, *24*, 1573–1583. [CrossRef]
22. Pohl, J.H.; Sarofim, A.F. Devolatilization and oxidation of coal nitrogen. *Symp. (Int.) Combust.* **1977**, *16*, 491–501. [CrossRef]
23. Kitto, J.B.; Stultz, S.C. *Steam—Its Generation and Use*, 41th ed.; Babcock & Wilcox Company: Charlotte, NC, USA, 2005.
24. Li, H.; Yoshihiko, N.; Dong, Z.; Zhang, M. Application of the FactSage to Predict the Ash Melting Behavior in Reducing Conditions. *Chin. J. Chem. Eng.* **2006**, *14*, 784–789. [CrossRef]
25. Wang, F.; Shen, B.; Yang, J.; Singh, S. Review of Mercury Formation and Capture from CO_2-Enriched Oxy-Fuel Combustion Flue Gas. *Energy Fuels* **2017**, *31*, 1053–1064. [CrossRef]
26. Yin, C.; Yan, J. Oxy-fuel combustion of pulverized fuels: Combustion fundamentals and modeling. *Appl. Energy* **2016**, *162*, 742–762. [CrossRef]
27. Xu, M.X.; Li, S.Y.; Li, W.; Lu, Q.G. Effects of gas staging on the NO emission during O_2/CO_2 combustion with high oxygen concentration in circulating fluidized bed. *Energy Fuels* **2015**, *29*, 3302–3311.
28. Scheffknecht, G.; Al-Makhadmeh, L.; Schnell, U.; Maier, J. Oxy-fuel coal combustion—A review of the current state-of-the-art. *Int. J. Greenh. Gas Control.* **2011**, *5*, S16–S35. [CrossRef]
29. Jankowska, S.; Czakiert, T.; Krawczyk, G.; Borecki, P.; Jesionowski, L.; Nowak, W. The Effect of Oxygen Staging on Nitrogen Conversion in Oxy-Fuel CFB Environment. *Chem. Process. Eng.* **2014**, *35*, 489–496. [CrossRef]
30. Watanabe, H.; Yamamoto, J.I.; Okazaki, K. NOx formation and reduction mechanisms in staged O_2/CO_2 combustion. *Combust. Flame* **2011**, *158*, 1255–1263. [CrossRef]
31. Gimenez-Lopez, J.; Millera, A.; Bilbao, R.; Alzueta, M.U. HCN oxidation in an O_2/CO_2 atmosphere: An experimental and kinetic modeling study. *Combust. Flame* **2010**, *157*, 267–276. [CrossRef]
32. Mendiara, T.; Glarborg, P. Ammonia chemistry in oxy-fuel combustion of methane. *Combust. Flame* **2009**, *156*, 1937–1949. [CrossRef]
33. Skjøth-Rasmussen, M.; Glarborg, P.; Østberg, M.; Johannessen, J.; Livbjerg, H.; Jensen, A.D.; Christensen, T. Formation of polycyclic aromatic hydrocarbons and soot in fuel-rich oxidation of methane in a laminar flow reactor. *Combust. Flame* **2004**, *136*, 91–128. [CrossRef]
34. Dagaut, P.; LeComte, F. Experiments and Kinetic Modeling Study of NO-Reburning by Gases from Biomass Pyrolysis in a JSR. *Energy Fuels* **2003**, *17*, 608–613. [CrossRef]
35. Zhang, X.; Sun, S.; Sun, R.; Li, X.; Zeng, G. Effects of bias combustion on volatile nitrogen transformation. *Asia-Pac. J. Chem. Eng.* **2009**, *5*, 473–478. [CrossRef]
36. Mendiara, T.; Glarborg, P. Reburn Chemistry in Oxy-fuel Combustion of Methane. *Energy Fuels* **2009**, *23*, 3565–3572. [CrossRef]
37. Glarborg, P.; Bentzen, L.L. Chemical Effects of a High CO_2 Concentration in Oxy-Fuel Combustion of Methane. *Energy Fuels* **2008**, *22*, 291–296. [CrossRef]
38. Wang, X.; Tan, H.; Niu, Y.; Chen, E.; Xu, T. Kinetic investigation of the SO_2 influence on NO reduction processes during methane reburning. *Asia-Pac. J. Chem. Eng.* **2010**, *5*, 902–908. [CrossRef]
39. Zajemska, M.; Musiał, D.; Poskart, A. Application of CHEMKIN and COMSOL Programs in the Calculations of Chemical Composition of Natural Gas Combustion Products. *Combust. Sci. Technol.* **2014**, *186*, 153–172. [CrossRef]

40. Lee, D.; Park, J.; Jin, J.; Lee, M. A simulation for prediction of nitrogen oxide emissions in lean premixed combustor. *J. Mech. Sci. Technol.* **2011**, *25*, 1871–1878. [CrossRef]
41. Hill, S.; Smoot, L.D. Modeling of nitrogen oxides formation and destruction in combustion systems. *Prog. Energy Combust. Sci.* **2000**, *26*, 417–458. [CrossRef]
42. Sanders, W.A.; Lin, C.Y.; Lin, M.C. On the Importance of the Reaction $CH_2 + N_2$? HCN+NH as a Precursor for Prompt NO Formation. *Combust. Sci. Technol.* **1987**, *51*, 103–108. [CrossRef]
43. Smith, G.P.; Golden, D.M.; Frenklach, M.; Moriarty, N.W.; Eiteneer, B.; Goldenberg, M.; Bowman, C.T.; Hanson, R.K.; Song, S.; Gardiner, W.C., Jr.; et al. GRI-Mech. Available online: http://www.me.berkeley.edu/gri_mech/ (accessed on 9 July 2020).
44. Miller, J.A.; Bowman, C.T. Mechanism and modeling of nitrogen chemistry in combustion. *Prog. Energy Combust. Sci.* **1989**, *15*, 287–338. [CrossRef]
45. Ślefarski, R. Study on the Combustion Process of Premixed Methane Flames with CO_2 Dilution at Elevated Pressures. *Energies* **2019**, *12*, 348. [CrossRef]

Communication

Synergy of Thermochemical Treatment of Dried Distillers Grains with Solubles with Bioethanol Production for Increased Sustainability and Profitability

Samuel O'Brien [1], Jacek A. Koziel [1,*], Chumki Banik [1] and Andrzej Białowiec [1,2]

1 Department of Agricultural and Biosystems Engineering, Iowa State University, Ames, IA 50011, USA; scobrien@iastate.edu (S.O.); cbanik@iastate.edu (C.B.); andrzej.bialowiec@upwr.edu.pl (A.B.)

2 Institute of Agricultural Engineering, The Faculty of Life Sciences and Technology, Wrocław University of Environmental and Life Sciences, 37a Chełmońskiego Str., 51-630 Wrocław, Poland

* Correspondence: koziel@iastate.edu; Tel.: +1-515-294-4206

Received: 7 July 2020; Accepted: 30 August 2020; Published: 1 September 2020

Abstract: The bioethanol industry continues improving sustainability, specifically focused on plant energy and GHG emission management. Dried distiller grains with solubles (DDGS) is a byproduct of ethanol fermentation and is used for animal feed. DDGS is a relatively low-value bulk product that decays, causes odor, and is challenging to manage. The aim of this research was to find an alternative, value-added-type concept for DDGS utilization. Specifically, we aimed to explore the techno-economic feasibility of torrefaction, i.e., a thermochemical treatment of DDGS requiring low energy input, less sophisticated equipment, and resulting in fuel-quality biochar. Therefore, we developed a research model that addresses both bioethanol production sustainability and profitability due to synergy with the torrefaction of DDGS and using produced biochar as marketable fuel for the plant. Our experiments showed that DDGS-based biochar (CSF—carbonized solid fuel) lower calorific value may reach up to 27 MJ·kg^{-1} d.m. (dry matter) Specific research questions addressed were: What monetary profits and operational cost reductions could be expected from valorizing DDGS as a source of marketable biorenewable energy, which may be used for bioethanol production plant's demand? What environmental and financial benefits could be expected from valorizing DDGS to biochar and its reuse for natural gas substitution? Modeling indicated that the valorized CSF could be produced and used as a source of energy for the bioethanol production plant. The use of heat generated from CSF incineration supplies the entire heat demand of the torrefaction unit and the heat demand of bioethanol production (15–30% of the mass of CSF and depending on the lower heating value (LHV) of the CSF produced). The excess of 70–85% of the CSF produced has the potential to be marketed for energetic, agricultural, and other applications. Preliminary results show the relationship between the reduction of the environmental footprint (~24% reduction in CO_2 emissions) with the introduction of comprehensive on-site valorization of DDGS. The application of DDGS torrefaction and CSF recycling may be a source of the new, more valuable revenues and bring new perspectives to the bioethanol industry to be more sustainable and profitable, including during the COVID-19 pandemic and other shocks to market conditions.

Keywords: biofuel; biorenewables; corn; DDGS; ethanol; sustainability; torrefaction; waste-to-energy; waste-to-carbon; CSF

---

1. Introduction

The production of bioethanol is a heat-consuming process. The heat demand for processing of corn is ~1100 MJ·Mg^{-1} (corn), while the production of bioethanol requires 38,600 MJ·Mg^{-1} (bioethanol).

Typically, this heat demand is supplied by the incineration of natural gas, a fossil fuel [1]. Therefore, the production of bioethanol is not considered sustainable (e.g., due to net GHG production, [2]), which is dependent on fluctuating prices of natural gas and subsidies. The bioethanol production industry needs continued efforts to be profitable, sustainable, self-sufficient, and independent from the volatile fuels market and resilient to changing policies.

We propose, for the first time, implementation of the new solution based on valorization and waste recycling according to 'waste to carbon' and 'waste to energy' approaches. One of the bioethanol industry waste is dried distiller grains with solubles (DDGS) accounting for ~30% [1] of the initial mass of corn. Wet DDGS is usually used for animal feeding with relatively low revenues for the ethanol plant. The wet (unprocessed) DDGS are prone to decay and are difficult to store without drying (and additional energy input for drying is required) [3].

We propose the use of torrefaction (roasting at 200 to 600 °C in the absence of oxygen) of DDGS to produce biochar. The biochar (i.e., carbonized DDGS, also known as carbonized solid fuel, CSF [4]) is then used as a fuel to that covers the entire heat needs for a bioethanol plant. The biochar produced from DDGS may reach the calorific value up to ~27 $MJ \cdot kg^{-1}$ d.m. (equivalent of high-quality coal, anthracite) while the mass yield of the biochar is 60–90%.

The proposed solution is an innovative change in the bioethanol production scheme due to: (i) addition of a torrefaction unit for thermochemical treatment of DDGS and production of biochar (CSF) characterized by improved fuel properties; (ii) substitution of CSF for natural gas (currently used as the heat source in a gas boiler) and incineration in a solid fuel boiler; (iii) use of heat generated from CSF incineration in solid fuel boiler for covering entire heat demands of torrefaction unit and the heat demand of bioethanol production process. The residual mass of CSF produced has the potential to be marketed for energetic, agricultural, and other applications.

In this paper, we present a techno-economic analysis showing that the proposed valorization of DDGS is profitable, i.e., it makes the bioethanol plant energy-neutral (for heat) while creating additional revenue streams from CSF sold outside. A few techno-economic models for bioethanol plants were developed. Kwiatkowski et al. created a model of the ethanol production system through the dry-grind process [1]. The model was created with the goal of being applied to the evaluation of current and new grain conversion technologies, to assist in the determination of how impactful alternative feedstocks can be, and for sensitivity analysis of high impact economic factors. As such, the model was not based on generic plant design. The model uses data from ethanol producers, manufacturers of the equipment, suppliers, and engineers in this field. The Kwiatkowski et al. [1] model was created to simply be a baseline for a standard ethanol production system.

Kalyian et al. [2] developed a model of a process that integrated torrefied corn stover into the ethanol production system. The goal of this model was to assess whether there was the potential to reduce greenhouse gas emissions through the integration of biomass torrefaction into industrial-scale businesses. Our model has a different scope in comparison to Kalyian et al. [2] because the biomass we propose as a feedstock for torrefaction is a direct byproduct of ethanol production. We are using DDGS for biochar production, which, in turn, could potentially earn positive carbon abatement effects.

We aimed to explore the techno-economic feasibility of torrefaction, i.e., a thermochemical treatment of DDGS requiring low energy input, less sophisticated equipment, and resulting in fuel-quality biochar. Therefore, we developed a research model that addresses both bioethanol production sustainability and profitability due to synergy with the torrefaction of DDGS and using produced biochar as marketable fuel for the plant. Specific research questions addressed were:

1. What monetary profits and operational cost reductions could be expected from valorizing DDGS as a source of marketable biorenewable energy?
2. What fraction of energy demand for bioethanol production could be supplied by CSF from DDGS?
3. What environmental and financial benefits could be expected from valorizing DDGS to CSF and its reuse for natural gas substitution?

2. Materials and Methods

2.1. Techno-Economic Model

A techno-economic model was created to provide an initial evaluation of the proposed concept. The model was based on data collected regarding ethanol production requirements and products [1]. Using these data, we were able to create a working model for a standard ethanol production system that can work with some key inputs such as the mass of corn being used, or the capacity of the plant, to generate information on the products and byproducts that would be produced as a result. What we used this information for was to create a hands-on model of our own so that we could compare a generic ethanol production system with our own proposed ethanol production system. The data in this paper was critical in the creation of our mass and energy balances, which were used to determine the viability of our proposed method. This whole model was created in Microsoft Excel (Supplementary Materials, Technoeconomic Model.xlsx).

The document itself contains several tabs providing information on the system and what is occurring in each step of the process. Among these tabs is a flowchart that provides not only a visual representation of the ethanol production system but also a detailed look at the mass breakdown at every step. A more detailed overview of what each tab contains can be seen in Table 1.

To make the model more user-friendly and organized, we used different colors to symbolize different processes in the flowchart. For example, the corn cleaning process is highlighted in Figure 1. The 'Total Corn' block is colored green as it is an input into the system, the 'Cleaning' block is colored blue to signify that it is a process in the system, and the 'Trash' block is red to signify a byproduct of the system that is not profitable. The only block not pictured is that of a profitable byproduct, which is colored in orange. On top of the color-coded blocks, several cells are colored yellow or orange, which yellow indicating values that are inputted and orange, indicating values that change as a result of the values in yellow.

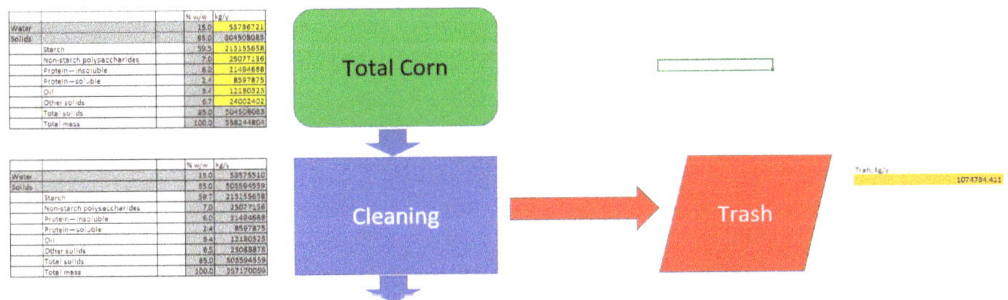

Figure 1. The cleaning process from the 'Flowchart' page in the Supplementary Materials (Excel spreadsheet, Technoeconomic Model).

An example of the techno-economic model is given in Figure 2, presenting the part of the mass balance of the bioethanol plant.

The developed techno-economic model was a source of the mass balance data for the simplified economic analysis of bioethanol plants. For calculations, we used four main components of the bioethanol plant:

- Corn prices as a cost.
- Natural gas prices as a cost.
- Ethanol prices as a revenue.
- DDGS prices as a revenue.

Additionally, in the scenario with the application of the torrefaction, we included the option of selling the CSF on the market. We assumed that the price of revenue would be similar to the price of biochar available as a soil amendment. We determined the CSF threshold price above which the bioethanol plant with implemented DDGS torrefaction starts to have higher profits than the conventional one.

Table 1. Legend to the Technoeconomic Model (Excel spreadsheet in Supplementary Materials): Bioethanol Plant Production Model.

Excel Tab (Name)	Function	Inputs	Outputs
Introduction	Brief explanation of what the file contains	NA	NA
Flowchart	Visual representation of ethanol production with the breakdown of the mass.	NA	• Mass of trash • Mass of ethanol • Mass of DDGS • Mass of CO_2
Substrate-corn	Provides information regarding the mass composition of corn.	• Yearly plant working hours • Annual cost of corn • Current corn price • Plant capacity	NA
Additives	Summary of the mass of all additives during the process.	• Water use	NA
Cleaning	Detailed description of the cleaning stage.	NA	NA
Grinding	Brief description of grinding and its relation to the mass balance.	NA	NA
Weigh Tank	Brief description of the weigh tank phase and its relation to the mass balance.	NA	NA
Slurry Tank	Breakdown of how the mass changes into a slurry and its resultant composition.	• Mass of α-amylase • Ammonia • Lime • Water	NA
Liquefaction	Brief overview of the liquefaction step and the conversation of the starch to oligosaccharides.	NA	NA
Saccharification	Brief overview of the saccharification step and the conversation of oligosaccharides to glucose.	NA	NA
Fermentation	Detailed look at the fermentation step and conversion rates to ethanol and biomass.	NA	NA
Beer Column	Brief overview of the routes the mass takes from the beer column.	NA	NA
Centrifugation	Description of centrifugation and assumptions for centrifuge capacity.	NA	NA
Ethanol Purification	Overview of the ethanol purification process as well as steam use.	NA	• Ethanol produced
Evaporation	Mass of water evaporated during purification.	NA	NA

NA—not applicable.

In our model, we used the most recent prices and costs given by the Hofstrand [5], which were converted to the relative price to a ton of corn. In our economic model, we included following components: Materials (Corn, Denaturant, Enzymes, Yeast, Chemicals, Other); Utilities (Electricity, Water, Natural Gas); Fixed costs (Labor and Management, Repairs and Maintenance, Property Taxes, Depreciation, Transportation, Interest); and Ethanol, DDGS, CSF.

All pricing data are given in Tables 2 and 3. The current price of biochar has been used to characterize the economic value of the CSF [6]. On the base of price data and mass and energy balances, we created the simplified economic model of these scenarios, which is available in the Supplementary Materials as an Economic analysis.xlsx file.

Cleaning:
During cleaning other (mineral, sand, dust, stones) solids are removed.
Rate of trash production, %
0.3

Annual production of trash, kg/y
1074734

Water	Removed	Residual
53736721	161210	53575510
Solids, kg/y	Removed other solids	Residual solids
304508083	913524	303594559
Other solids	Removed other solids	Residual other solids
24002402	913524	23088878

Figure 2. Information provided in the 'Cleaning' page from the Supplementary Material (Excel spreadsheet, Technoeconomic Model).

Table 2. The economic analysis of the conventional bioethanol plant on the basis of the mass balance of the main components and prices.

Bioethanol Plant Model Component	Reference	The Price Per Mass Equivalent of Model Component, $\$\cdot ton^{-1}$	The Price Per Mass Equivalent of Corn of the Model Component, $\$\cdot ton^{-1}$ Corn	Mass of Model Component, $t\cdot y^{-1}$	Costs, $\$\cdot y^{-1}$	Revenues, $\$\cdot y^{-1}$
			Materials			
Corn	[1,5]	122	-	358,250	43,705,866	-
Denaturant	[5]	-	5.14	-	1,840,306	-
Enzymes	[5]	-	3.69	-	1,322,967	-
Yeast	[5]	-	0.37	-	134,271	-
Chemicals	[5]	-	3.57	-	1,279,526	-
Other	[5]	-	2.20	-	789,831	-
			Utilities			
Electricity	[5]	-	4.33	-	1,552,736	-
Water	[5]	-	1.35	-	483,772	-
Natural Gas	[1,5]	42.4	-	11,150	472,760	-
			Fixed costs			
Labor and Management	[5]	-	4.19	-	1,502,115	-
Repairs and Maintenance	[5]	-	2.76	-	987,289	-
Property Taxes	[5]	-	0.25	-	89,754	-
Depreciation	[5]	-	4.67	-	1,674,442	-
Transportation	[5]	-	0.83	-	2,96,187	-
Interest	[5]	-	6.99	-	2,502,742	-
Subtotal	-	-	-	-	58,634,563	
			Income			
Ethanol	[1,5]	398	-	124,000	-	49,344,040
DDGS	[1,5]	127	-	114,200	-	14,505,432
Profit			5,214,909			

Calculation of Energy Demand for Torrefaction of DDGS

The results from the DSC analyses were used to calculate the actual energy demand in processing dry DDGS (to heat DDGS from 20 °C to 260 °C) according to the procedure described by Świechowski et al. [7]. It has been assumed that DDGS is dried, and energy is not spent on water evaporation. The energy demand for DDGS drying is included in the natural gas demand of bioethanol plant. The energy needed to DDGS torrefaction was calculated by Equation (1):

$$Q = m \cdot \Delta T \cdot cp \tag{1}$$

where:

Q—the total amount of heat needed to heat DDGS, GJ·y^{-1},
m—the mass of the DDGS, t·y^{-1},
ΔT—the temperature difference between ambient temperature (20 °C) and torrefaction point (260 °C), under normal pressure conditions, °C,
cp—specific heat of DDGS, 1.6 kJ·(kg·°C)$^{-1}$.

2.2. Production of CSF

Corn DDGS originated from commercial provider [8] "Goświnowice" Ethanol Plant Głębinów, Poland. DDGS torrefaction was done according to Syguła et al. [9] in a muffle furnace (SNOL 8.1/1100, Utena, Lithuania). CO_2, with a flow rate of 10 dm^3·h^{-1}, was supplied to create oxygen-free conditions. The process was carried out under temperature 260 °C during 60 min retention time. The samples were heated with a heating rate of 50 °C·min^{-1} from 20 °C to the torrefaction setpoint temperature. Ten grams (±0.5 g) of the dry mass of the sample was placed in the steel crucible. The CSF was removed from the muffle furnace when the interior temperature was lower than 200 °C (temperature was monitored by an internal thermocouple and visualized on the screen of the furnace). The approximate times of cooling from 260 °C to 200 °C was 29 min, respectively.

Table 3. The economic analysis of the bioethanol plant with implemented torrefaction of DDGS on the base of the mass balance of the main components and prices of the main component of the techno-economic model of bioethanol plant.

Bioethanol Plant Model Component	Reference	The Price Per Mass Equivalent of Model Component, \$·ton^{-1}	The Price Per Mass Equivalent of Corn of the Model Component, \$·ton^{-1} Corn	Mass of Model Component, t·y^{-1}	Costs, \$·y^{-1}	Revenues, \$·y^{-1}
			Materials			
Corn	[1,5]	122	-	358,250	43,705,866	-
Denaturant	[5]	-	5.14	-	1,840,306	-
Enzymes	[5]	-	3.69	-	1,322,967	-
Yeast	[5]	-	0.37	-	134,271	-
Chemicals	[5]	-	3.57	-	1,279,526	-
Other	[5]	-	2.20	-	789,831	-
			Utilities			
Electricity	[5]	-	4.33	-	1,552,736	-
Water	[5]	-	1.35	-	483,772	-
Natural Gas	[1,5]	42.4	-	0	0	-
			Fixed costs			
Labor and Management	[5]	-	4.19	-	1,502,115	-
Repairs and Maintenance	[5]	-	2.76	-	987,289	-
Property Taxes	[5]	-	0.25	-	89,754	-
Depreciation	[5]	-	4.67	-	1,674,442	-

Table 3. *Cont.*

Bioethanol Plant Model Component	Reference	The Price Per Mass Equivalent of Model Component, $·ton^{-1}	The Price Per Mass Equivalent of Corn of the Model Component, $·ton^{-1} Corn	Mass of Model Component, t·y^{-1}	Costs, $·y^{-1}	Revenues, $·y^{-1}
Transportation	[5]	-	0.83	-	2,96,187	-
Interest	[5]	-	6.99	-	2,502,742	-
Subtotal	-	-	-	-	58,634,563	-
			Income			
Ethanol	[1,5]	398	-	124,000	-	49,344,040
CSF	[10]	2614	-	74,462	-	194,643,668
Profit			185,825,905			

2.3. DDGS and CSF Analyses

The DDGS and CSF samples were tested in three replicates of high heating value (HHV) and lower heating value (LHV), determined in accordance with [10], by means of the IKA C2000 calorimeter. The dried samples of raw DDGS were tested by differential scanning calorimetry (DSC) analyses, according to procedure and equipment described by Świechowski et al. [4] and the DDGS heat capacity (cp, kJ·(kg·°C)$^{-1}$) was determined. The mass yield, energy densification ratio, and energy yield of CSF were determined based on Equations (2)–(4) [11], respectively:

$$MY = m_b/m_a \cdot 100 \tag{2}$$

where MY is the mass yield (%), m_a is the mass of raw material before torrefaction (kg), and m_b is the mass of CSF after torrefaction (kg).

$$EDr = HHV_b/HHV_a \tag{3}$$

where EDr is the energy densification ratio, $HHVb$ is the high heating value of CSF (MJ·kg^{-1}), and HHV_a is the high heating value of raw DDGS (MJ·kg^{-1}).

$$EY = MY \cdot EDr \tag{4}$$

where EY is the energy yield (%), MY is the mass yield (%), and EDr is the energy densification ratio.

3. Results

3.1. CSF Properties

The DDGS torrefaction process carried under 260 °C and for 60 min was characterized by high biomass conversion efficiency. The mean HHV and LHV of DDGS were 20.76 ± 0.45 and 19.53 ± 0.38 MJ·kg^{-1} d.m., respectively, while CSF was characterized by HHV and LHV at 27.03 ± 0.31 and 26.12 ± 0.27 MJ J·kg^{-1} d.m., respectively. The energy densification ratio was 1.3. The mass yield was 77.3 ± 0.7%. In total, the energy yield reached 100.6%.

3.2. Conventional Bioethanol Plant Techno-Economic Model

The schematic below shows the simplified technological process of bioethanol production from corn. It requires natural gas (for heat) and generates waste (a highly unstable, difficult to manage and store byproduct, DDGS) (Figure 3).

Based on the model, ~358,245 tons of corn must be utilized per year to produce 123,984 tons per year of denatured bioethanol. The corn is the main cost, accounting for ~$44M·y^{-1} (Table 2). The bioethanol plant operation requires ~11,149 tons of natural gas to be utilized per year [1]. The cost of natural gas, according to the current gas prices (Table 2), was ~$473K·y^{-1}. The total costs of the bioethanol plant production were ~$58.6M·y^{-1}. The revenues from bioethanol was ~$49M·y^{-1} (Supplementary Materials,

Economic analysis.xlsx, Table 2). Except for the bioethanol, DDGS is produced in the amount of 114,216 ton per year. Usually, it is consumed as an animal feed. According to current DDGS market prices, the revenue value from selling the DDGS is ~\$14.5M·y^{-1} (Table 2, Supplementary Materials, Economic analysis.xlsx). Our calculations indicated that traditional bioethanol plant profit was about \$5M·y^{-1} (Table 2, Supplementary Materials, Economic analysis.xlsx).

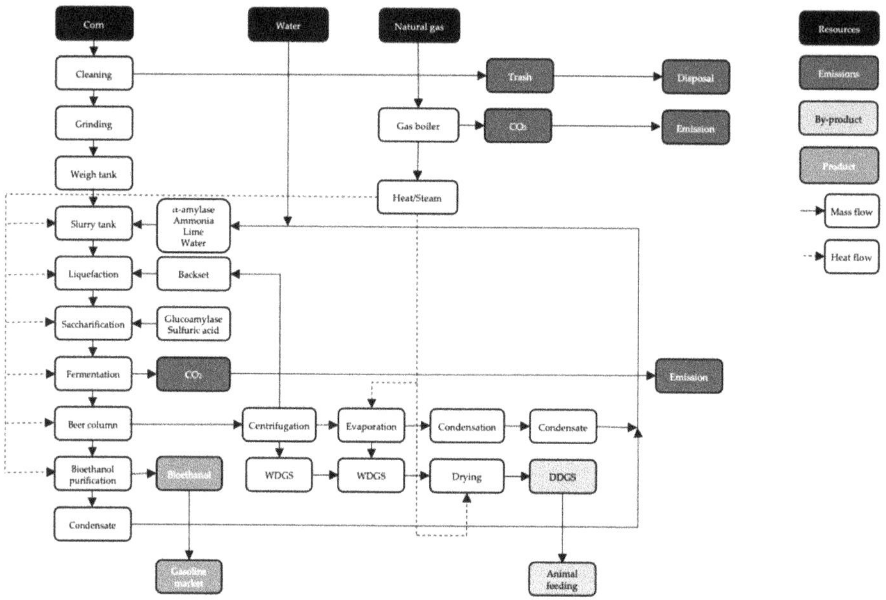

Figure 3. A detailed breakdown of the processes in a conventional ethanol plant.

Concerning the environmental issues related to bioethanol production, the total emission of CO_2 127,884 t CO_2·y^{-1}, consisting of the sum of natural gas use (30,659 t CO_2·y^{-1}) and fermentation process (97,225 t CO_2·y^{-1}) (Supplementary Materials, Technoeconomic Model.xlsx).

3.3. *Innovative Bioethanol Plant with the CSF Production Techno-Economic Model*

We proposed to valorize DDGS by making fuel from it. Thus, the plant will become energy-neutral by eliminating natural gas burning. An energy-neutral plant does not mean that there is zero additional energy added to the system but rather that it is low and more sustainable, which is what our system proposes. Usually, the bioethanol plant does not have its own electricity production unit and imports the electricity from the grid. The new fuel produced (CSF) can be sold outside and create a new revenue stream. Therefore, we proposed to apply the torrefaction process of DDGS, resulting in the production of CSF, which can be reused for heating the bioethanol plant—'heating' for saccharification, liquefaction, fermentation, ethanol distillation and WDGS drying (12.6% of CSF produced), and for 'heating' the torrefaction process (3.1% of CSF produced). The remainder 84.3% of CSF may be sold on the solid fuel market and be a new source of revenues (Figure 4).

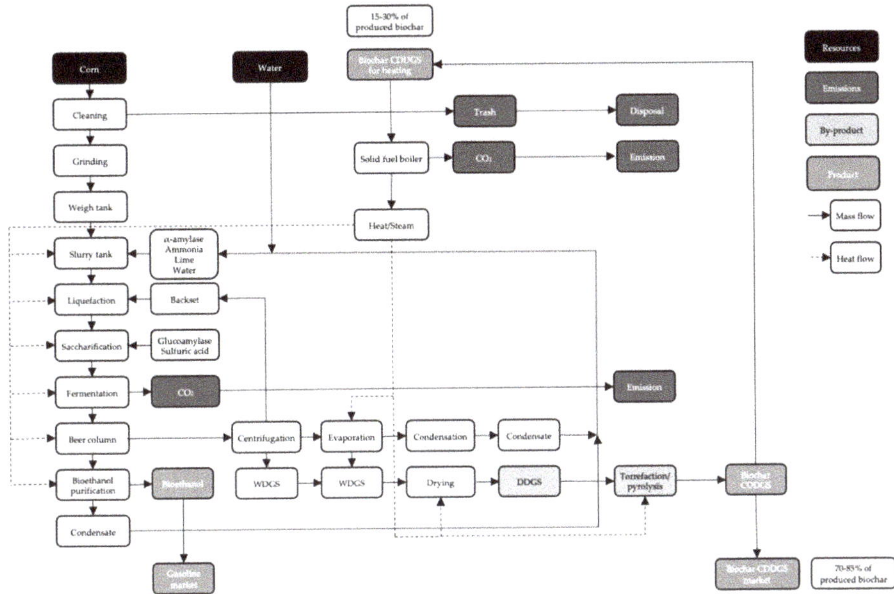

Figure 4. A detailed breakdown of the processes in the proposed ethanol production system.

Accounting the mass yield of DDGS torrefaction 77.3%, the production of CSF should reach 88,289 t per year. The torrefaction of one ton of DDGS requires 624 MJ. It means that for the total mass of DDGS, the energy demand on torrefaction is 71,270,784 MJ·y^{-1}. Taking LHV of CSF to be 26.12 MJ·kg^{-1} d.m. to cover the DDGS energy demand, ~2729 t·y^{-1} of CSF must be utilized, i.e., as low as ~3.1% of the total CSF production. Knowing that the bioethanol plant natural gas demand is 11,149 t·y^{-1} and that the LHV of natural gas is ~26 MJ·kg^{-1}, ~11,098 t/year of CSF must be used to substitute the natural gas, i.e., as low as ~12.6% of the total CSF production. Moreover, the substitution of natural gas by CSF reduces CO_2 emission by 30,659 t of CO_2·y^{-1} (about 24% of the total CO_2 emission). Considering that CSF is a 'green' renewable fuel, the CO_2 emission from CSF burning should not be included in the CO_2 balance. These calculations showed that ~84% (~74,462 t·y^{-1}) of CSF might be sold as a high-quality fuel. An additional benefit is that the savings from the natural gas substitution by CSF reach ~$472K·y^{-1} (Table 3, Supplementary Materials, Economic analysis.xlsx).

Due to the very high biochar price used for calculations, the overall profit of the bioethanol plant exceeded $185M.y^{-1}, which is probably not realistic. Therefore, we estimated that the minimum price for CSF above which the bioethanol plant has a similar level of profit as the conventional one is about $188·t^{-1} (Supplementary Materials, Economic analysis.xlsx). The current market price for biochar is an order of magnitude higher, exceeding $2600·t^{-1} [5]. It shows that new products such as CSF may allow the replacement of natural gas and the creation of new, value-added products giving the perspectives to be more profitable, therefore making the bioethanol industry more sustainable.

4. Discussion

Overall, the proposed solution increases the bioethanol production sustainability and profitability due to synergy with the thermochemical treatment of DDGS (via torrefaction), reuse of part of the produced CDDGS for covering all plant heat demand, replacing the need to purchase natural gas (fuel), and creating a new revenue stream for a plant. The majority of biochar produced from DDGS can be sold, and thus, be considered as a new revenue stream for a bioethanol plant.

Ethanol plants are, therefore, faced with the question if the existing production line of the bioethanol plant should be modified to include the DDGS torrefaction unit. The bioethanol industry is a well-developed branch of the economy. However, it suffers from a frequent negative return rate over the operating costs (Figure 5). That situation has been made worse by the COVID-19 pandemic, when both bioethanol and corn prices dropped. Such a situation, when the existing system is suffering from the significant drop of bioethanol and corn prices (lower revenues both for bioethanol plant and farmers), is an opportunity for the implementation of innovations that under normal conditions would not be considered by decisionmakers. There is also a question if such an innovative solution should be integrated with the bioethanol plan or should be a separate unit.

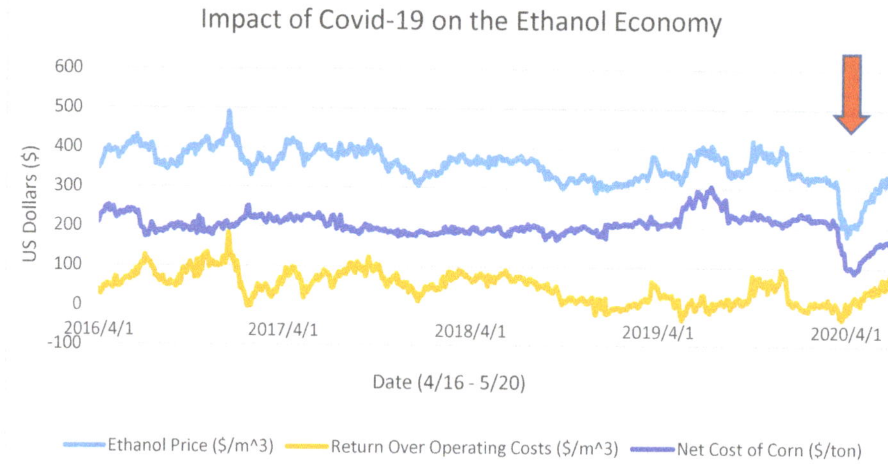

Figure 5. Prices of ethanol (2015–2020) [12]. A red arrow indicates an example of the recent price drop caused by the SARS-Cov-2 outbreak.

Haseli [13] created a model that can be used to provide a rough estimate of what the optimum operating point and energy consumption is for a torrefaction plant. The goal was to create a tool that could provide information on how to maximize net energy gain in a standalone and integrated model. The data used was based on the study of two different types of wood, which is run in both scenarios and used to determine what is the ideal energy consumption to produce the most mass of torrefied material. Haseli [13] model of torrefaction integration with a power plant, suggest a downstream method that requires a separate plant for the torrefaction. We are not proposing a downstream model like Haseli [13]; instead, we are proposing the direct modification of the original process of the dry-grind ethanol production system, which would directly loop the torrefaction of the biological material into the system. Therefore, by the implementation and integration of the torrefaction process with bioethanol plants, there is a synergistic effect causing the increase of economic efficiency. Our model includes previous research from our team on finding the ideal conditions for torrefaction, using DDGS instead of wood, which would maximize the mass yield and calorific value of DDGS.

Kalyian et al. [2] developed a model of a process that integrated torrefied corn stover into the ethanol production system. The goal was to assess whether there was the potential to reduce greenhouse gas emissions through the integration of biomass torrefaction into industrial-scale businesses. Based on the mass of corn stover collected in his model, which was 70% of corn stover per unit land area on a biannual basis, it would be able to meet 42.8% of the steam demands of a dry-grind corn ethanol plant in the US. It was found that this would result in a 40% reduction of greenhouse gas emission for corn-based ethanol when compared to gasoline. A paper by Kalyian et al. [2] has a primary focus on the environmental impacts' corn stover torrefaction provides.

Our idea differs from Kalyian et al. [2] in many instances. First and foremost is the choice of biomass. Corn stover is known to be left in field post-harvest for the positive impact it has on the soil, and this is addressed in [2]. Our choice of biomass (DDGS) would not be associated with concerns about amending current agronomic practices that can have a comparably negative impact on soil. In addition, we have found that with the amount of biochar produced from DDGS, we can cover the total energy demands of an ethanol plant while maintaining a significant percentage of biochar left over to be sold.

One of the challenges of the bioeconomy implementation is the overcoming of the problem of high carbon footprint, sustainability, and energy efficiency of the biorefineries. This issue opens a new niche for the ideas, interdisciplinary problem solving, and innovation in the wide field of the bioeconomy. Schipfer and Kranzl [14] addressed the European Union to shift to a biobased system as a means of reducing emissions and fossil fuel usage. The techno-economic evaluation of biomass to the product uses two feedstocks, i.e., wood chips (beechwood) and wheat straw. This paper addressed the possibility of integrating bioenergy into residential and industrial scales. Ortiz et al. [15] performed a review of three different models for sugarcane-based biorefineries. The three options were a conventional biorefinery, a biochemical platform that utilizes bagasse as a fuel for the system, and finally, a thermochemical platform that utilizes gasification. Kumar et al. [16] addressed the differences between the methods of producing conventional biochar (made from dry feedstock) and hydrochar (made from wet feedstock). The generation of conventional biochar used pyrolysis, torrefaction, flash carbonization, and gasification, whereas hydrothermal carbonization generates hydrochar. The findings indicate that a higher energy density can be expected largely due to the lower ash content in the produced hydrochar. In our study, we used conventional torrefaction, which has higher TRL (Technology Readiness Level), i.e., it has already been implemented at the full scale with wood [17], compared to the hydrothermal carbonization, i.e., still at the stage of R&D and prototyping. However, the next step could be the execution of tests leading to hydrochar production from DDGS, and modeling the influence on the bioethanol plant sustainability and profitability. Manouchehrinejad and Mani [18] used pine wood chips as a feedstock for simulating an integrated biomass torrefaction and pelletization (iBTP) plant for the production of solid biofuels. The simulation found that the 100,000 $Mg \cdot y^{-1}$ plant could become autothermal for torrefaction over 270 °C requiring 8.2~9 $MJ \cdot kg^{-1}$ for the production of torrefied pellets.

The proposed solution is an innovative creation of the new product biochar from DDGS, which could be called CSF (Carbonized Solid Fuel), with the potential to be marketed for energetic and agricultural applications.

The proposed invention may bring the following benefits:

(1) Elimination of natural gas purchase costs.
(2) A new source of revenue from selling the CSF.
(3) Improved the degree of independence from fluctuations in the raw materials market.
(4) Implementation of the 'circular economy' inside the bioethanol plant.

From a technical-economic point of view, the most important drivers for the development of the bio-based industry are the economic impact and process sustainability [19]. These drivers include the profitability of the company and the environmental performance of the products. Additionally, the policy may be considered a significant driver with a secondary role, although it may become more important in the future because policy acts at the end as a common channel for the demands of society and the strategies of companies. During bio-products generation (bioethanol), part of the biomass will be converted to waste stream—DDGS. This organic waste must be recycled. We propose, for the first time, the implementation of a new solution based on valorization and biowaste recycling according to the 'waste to carbon' and 'waste to energy' approaches. The wet organic waste is prone to decay and is difficult to store without drying (and additional energy input for drying is required). The concept of the closing the biomass supply chain due to the torrefaction of DDGS for CSF production, its reuse for the bioethanol plant heat demand covering, and its recycling to the agricultural or energetic purposes falls into the "Roadmap of Transformation to Circular Economy" adopted by the Polish Government in the 2019—a new policy regarding the circular economy, including the bioeconomy [20]. Torrefaction is a technology that can serve as a basis for the development of other technologies, by acting as a preprocessing technology prior to the use of other processes, whether for energy production or for biorefineries for green chemicals. The use of torrefaction as a pretreatment technology allows the gasification process to be much more efficient than when starting from thermally unprocessed biomass. The production of more advanced forms of materials with high fixed carbon contents, such as charcoal, coke, and activated charcoal, are also promising possibilities that will aid with the development of new nanotechnology products, for example, by supplying carbon for the production of graphene, or activated charcoal for the removal of toxic compounds [21].

Nowadays, the torrefaction of biomass is a developing technology. Still, due to the characteristics of the resulting products, it generates the interest of the sector's investors. From this perspective, the potential development of biomass thermal conversion technologies, such as torrefaction and/or carbonization, is considered promising regarding the utilization of new forms of biomass; namely, the less environmentally-friendly, more abundant, and faster-growing forms, as is the case of shrub plants [22] or DDGS as in the presented case. The torrefaction cannot yet be considered as a mature technology, thus, a significant investment in R&D is still needed [22], including the integration with bioethanol plant or other biorefineries.

We intended to present the proof-of-the-concept of such a synergistic combination of bioethanol plant with DDGS torrefaction and biochar recycling. However, in this paper, we included the economic model (Supplementary Materials, Economic analysis.xlsx) by inserting the process data related to the ton of corn, where potential users could input their data of the mass of corn, from their bioethanol plants to do the calculations of potential benefits from application of our idea. In the model, they can input data about the mass of corn, mass of ethanol, mass of natural gas, mass of DDGS, and mass of CSF. The model will calculate the final revenues considering all possible costs.

The presented model is the proof-of-the-concept, which should be further studied, including scaling up the system, integration of the devices, life cycle assessment, and final feasibility study, before the implementation.

5. Conclusions

The proposed solution is an innovative change in the bioethanol production scheme due to:

- Addition of a torrefaction unit for thermochemical treatment of DDGS and production of biochar (CSF) characterized by the lower calorific value (LCV) < ~30 $MJ \cdot kg^{-1}$ and >19 $MJ \cdot kg^{-1}$;
- Substitution of natural gas (currently used as a heat source in a gas boiler) with CSF and incineration in a solid fuel boiler. It can result in the monetary profits $\$473K \cdot y^{-1}$ due to savings of the cost of the natural gas that would no longer be needed. Depending on the market, when the price of the CSF is higher than $\$188 \cdot t^{-1}$, the bioethanol plant could start to make a profit at a higher rate than a conventional one.
- Use of heat generated from CSF incineration in solid fuel boiler for covering entire heat demands of torrefaction unit and the heat demand of bioethanol production line (15–30% of the mass of CSF and depending on LHV of CSF produced).
- In total, 70–85% of the CSF produced has the potential to be marketed for energetic, agricultural, and other applications.
- The substitution of natural gas by CSF reduces CO_2 emission by ~24% of the total CO_2 emission of the bioethanol plant.

- The application of DDGS torrefaction and CSF recycling may be a source of the new, more valuable revenues and bring new perspectives to the bioethanol industry to be more sustainable and profitable, including during the COVID-19 (or other diseases) pandemic and other extreme market conditions.

6. Patents

The early version of the work described in this manuscript was submitted to Iowa State University as an Invention Disclosure [23] Koziel, J.A., S. O'Brien, C. Banik, A. Bialowiec. Energy-neutral bioethanol production plant. Iowa State University. Invention Disclosure Docket No. 05017. (Submitted on 9 October 2019).

Supplementary Materials: The following are available online at http://www.mdpi.com/1996-1073/13/17/4528/s1, Technoeconomic Model.xlsx, and Economic analyses.xlsx.

Author Contributions: Conceptualization, A.B., S.O., and J.A.K.; methodology, A.B., S.O., J.A.K., and C.B.; software, S.O. and A.B.; validation, S.O. and A.B.; formal analysis, S.O., J.A.K., A.B., and C.B.; investigation, S.O. and A.B.; resources, A.B., S.O., J.A.K., and C.B.; data curation, A.B., S.O., and J.A.K.; writing—original draft preparation, S.O.; writing—review and editing, A.B., S.O., J.A.K., and C.B.; visualization, A.B., S.O., and J.A.K.; supervision, A.B. and J.A.K.; project administration, A.B. and J.A.K.; funding acquisition, A.B. and J.A.K. All authors have read and agreed to the published version of the manuscript.

Funding: The authors would like to thank the Fulbright Foundation for funding the project titled "Research on pollutants emission from Carbonized Refuse-Derived Fuel into the environment," completed at the Iowa State University. In addition, this paper preparation was partially supported by the Iowa Agriculture and Home Economics Experiment Station, Ames, Iowa. Project no. IOW05556 (Future Challenges in Animal Production Systems: Seeking Solutions through Focused Facilitation) sponsored by Hatch Act and State of Iowa funds.

Acknowledgments: The authors are thankful to the Iowa State University Honors Program for facilitating the research mentoring matching for Samuel O'Brien via the Honors First-Year Mentor Program. The presented article results were obtained as part of the activity of the leading research team—Waste and Biomass Valorization Group (WBVG), https://www.upwr.edu.pl/research/50121/waste_and_biomass_valorization_group_wbvg.html.

Conflicts of Interest: The authors declare no conflict of interest.

References

1. Kwiatkowski, J.R.; McAllon, A.J.; Taylor, F.; Johnston, D.B. Modeling the process and costs of fuel ethanol production by the corn dry-grind process. *Ind. Crops Prod.* **2006**, *23*, 288–296. [CrossRef]
2. Kaliyan, N.; Morey, R.V.; Tiffany, D.G.; Lee, W.F. Life cycle assessment of a corn stover torrefaction plant integrated with a corn ethanol plant and a coal fired power plant. *Biomass Bioenergy* **2014**, *63*, 92–100. [CrossRef]
3. Rosentrater, K.A. Ethanol processing coproducts—A review of some current constraints and potential directions. *Int. Sugar J.* **2007**, *109*, 1–12.
4. Świechowski, K.; Syguła, E.; Koziel, J.A.; Stępień, P.; Kugler, S.; Manczarski, P.; Białowiec, A. Low-temperature pyrolysis of municipal solid waste components and refuse-derived fuel—Process efficiency and fuel properties of carbonized solid fuel. *Data* **2020**, *5*, 48. [CrossRef]
5. Hofstrand, D. 07.01.2020. Available online: http://www.extension.iastate.edu/agdm/energy/xls/d1-10ethanolprofitability.xlsx (accessed on 19 August 2020).
6. ARTi. Available online: https://artichar.com (accessed on 27 June 2020).
7. Swiechowski, K.; Stepien, P.; Hnat, M.; Kugler, S.; Stegenta-Dabrowska, S.; Koziel, J.A.; Manczarski, P.; Bialowiec, A. Waste to carbon: Biocoal from elephant dung as new cooking fuel. *Energies* **2019**, *12*, 22.
8. Bioagra. Available online: https://bioagra.pl/en/products/ddgs-distillers-dry-grains-with-solubles/ (accessed on 25 June 2020).
9. Syguła, E.; Koziel, J.A.; Białowiec, A. Proof-of-concept of spent mushrooms compost Torrefaction—Studying the process kinetics and the influence of temperature and duration on the calorific value of the produced biocoal. *Energies* **2019**, *12*, 3060. [CrossRef]
10. Polish Committee for Standardization. Solid Biofuels—Determination of Calorific Value. PN-EN ISO 18125:2017-07, 12 July 2017.

11. Chin, K.L.; H'ng, P.S.; Go, W.Z.; Wong, W.Z.; Lim, T.W.; Maminski, M.; Paridah, M.T.; Luqman, A.C. Optimization of torrefaction conditions for high energy density solid biofuel from oil palm biomass and fast growing species available in Malaysia. *Ind. Crops Prod.* **2013**, *49*, 768–774. [CrossRef]
12. Center for Agricultural and Rural Development. Available online: https://www.card.iastate.edu/ (accessed on 23 June 2020).
13. Haseli, Y. Simplified model of torrefaction-grinding process integrated with a power plant. *Fuel Process. Technol.* **2019**, *188*, 118–128. [CrossRef]
14. Schipfer, F.; Kranzl, L. Techno-economic evaluation of biomass-to-end-use chains based on densified bioenergy carriers (dBECs). *Appl. Energy* **2019**, *239*, 715–724. [CrossRef]
15. Ortiz, P.A.S.; Maréchal, F.; de Oliveira Junior, S. Exergy assessment and techno-economic optimization of bioethanol production routes. *Fuel* **2020**, *279*, 118327. [CrossRef]
16. Kumar, A.; Saini, K.; Bhaskar, T. Review—Hydrochar and biochar: Production, physicochemical properties and technoeconomic analysis. *Bioresour. Technol.* **2020**, *310*, 123442. [CrossRef] [PubMed]
17. Thrän, D.; Witt, J.; Schaubach, K.; Kiel, J.; Carbo, M.; Maier, J.; Ndibe, C.; Koppejan, J.; Alakangas, E.; Majer, S.; et al. Moving torrefaction towards market introduction—Technical improvements and economic-environmental assessment along the overall torrefaction supply chain through the SECTOR project. *Biomass Bioenergy* **2016**, *89*, 184–200. [CrossRef]
18. Manouchehrinejad, M.; Mani, S. Process simulation of an integrated biomass torrefaction and pelletization (iBTP) plant to produce solid biofuels. *Energy Convers. Manag. X* **2019**, *1*, 100008. [CrossRef]
19. Nattrass, L.; Biggs, C.; Bauen, A.; Parisi, C.; Rodríguez-Cerezo, E.; Gómez-Barbero, M. The EU Bio-Based Industry: Results from a Survey. Joint Research Centre, 2016. Available online: http://publications.jrc.ec.europa.eu/repository/bitstream/JRC100357/jrc100357.pdf (accessed on 3 August 2020).
20. Prime Minister Office. Government of Poland. Resolution on the Adoption of the "Roadmap for Transformation towards a Circular Economy", 10 September 2019. Available online: https://www.premier.gov.pl/wydarzenia/decyzje-rzadu/uchwala-w-sprawie-przyjecia-mapy-drogowej-transformacji-w-kierunku.html) (accessed on 3 August 2020).
21. Nunes, L.J.R.; Matias, J.C.O. Biomass torrefaction as a key driver for the sustainable development and decarbonization of energy production. *Sustainability* **2020**, *12*, 922. [CrossRef]
22. Ribeiro, J.M.C.; Godina, R.; Matias, J.C.O.; Nunes, L.J.R. Future perspectives of biomass torrefaction: Review of the current state-of-the-art and research development. *Sustainability* **2018**, *10*, 2323. [CrossRef]
23. Koziel, J.A.; O'Brien, S.; Banik, C.; Bialowiec, A. Energy-Neutral Bioethanol Production Plant. Invention Disclosure Docket No. 05017, 9 October 2019.

MDPI
St. Alban-Anlage 66
4052 Basel
Switzerland
Tel. +41 61 683 77 34
Fax +41 61 302 89 18
www.mdpi.com

Energies Editorial Office
E-mail: energies@mdpi.com
www.mdpi.com/journal/energies

www.ingramcontent.com/pod-product-compliance
Lightning Source LLC
LaVergne TN
LVHW070654170726
843515LV00004B/403

* 9 7 8 3 0 3 6 5 0 3 2 2 6 *